THE COMETS
AND THEIR ORIGIN

THE COMETS
AND THEIR ORIGIN

BY

R. A. LYTTLETON

Fellow of St. John's College, Cambridge,
and University Lecturer in Mathematics

CAMBRIDGE
AT THE UNIVERSITY PRESS
1953

CAMBRIDGE UNIVERSITY PRESS
Cambridge, New York, Melbourne, Madrid, Cape Town,
Singapore, São Paulo, Delhi, Mexico City

Cambridge University Press
The Edinburgh Building, Cambridge CB2 8RU, UK

Published in the United States of America by Cambridge University Press, New York

www.cambridge.org
Information on this title: www.cambridge.org/9781107615618

First published 1953
First paperback edition 2013

A catalogue record for this publication is available from the British Library

ISBN 978-1-107-61561-8 Paperback

CONTENTS

PREFACE

During recent years the subject of Comets has received little attention by astronomers, apart from the routine work of observation and computation of orbits. The theory of their origin has been almost completely neglected (of necessity, in the absence of hypotheses), and the obscurity attaching to the whole subject of Comets as a cosmogonical problem had come to be accepted as yet another of the numerous mysteries of astronomy. It has been one of the principal successes of the New Cosmology that, without having any idea of an attack on the cometary problem in view, nevertheless one of the fundamental processes discovered in connexion with stellar evolution has been found to lead quite naturally to a straightforward, and indeed a necessary, explanation of the presence of comets in the solar system, and also leads on to an understanding of many of their properties. This book represents an attempt to lay this theory before as wide a circle of astronomers as possible, in the hope that it will bring about renewed interest in the subject of Comets and thereby help to integrate astronomical theory into a united philosophical whole instead of remaining a closely guarded patchwork of disconnected, more or less taxonomic descriptions.

It has seemed to me to be more than desirable to present also an account of the observational features of Comets, which do not appear to be by any means widely known, and this information I have culled from the vast literature of the subject. I have no direct observational experience, at any rate with a telescope, and I claim no originality for this material. I express my indebtedness to the numerous authors, most of whom are no longer with us, whose papers and writings I have found so absorbingly interesting; and I hope that too much of that element of interest has not disappeared as a result of my summarization and selection of their writings. The first two chapters of this book contain this account. There follow two chapters on the theory of the formation and structure of comets, and then by way of conclusion follows a short chapter showing the relation of the work to earlier attempts at theoretical explanations. An Appendix gives numerous references to the literature of Comets, but is not claimed to be exhaustive.

My special thanks are due to my colleagues, Mr F. Hoyle and Mr H. Bondi, of the Faculty of Mathematics at Cambridge, and Mr T. Gold, of the Cavendish Laboratory, for many helpful discussions. The inclusion of a number of photographs of the more celebrated Comets, though not essential to the study of the book, seemed to me appropriate, and I have to express my thanks to the Directors of the several observatories at which these were originally taken for their permission to include them. My thanks are also due to the Cambridge University Press for the unremitting care they have bestowed on the production of the book.

R. A. LYTTLETON

St John's College,
Cambridge
March, 1952

LIST OF PLATES

I

DYNAMICAL PROPERTIES OF COMETS

Introduction

Until the great advances of very recent years in almost every branch of theoretical astronomy, comets had come to be regarded as perhaps the most puzzling and mysterious of all the many types of heavenly bodies. This was not due to any great difficulty in the way of observing comets, as could be urged with a problem such as the structure of the solar corona, for they are extremely numerous, and some of them at times so extraordinarily bright as to outshine Venus herself and be easily visible in broad daylight with the naked eye. Their various properties are almost the direct antithesis of those of planets, and the two types of object are often referred to as 'the two solar families'. The times and intervals of occurrence of comets in the sky seem quite irregular compared with stars and planets, some remaining visible for months and others only for a few days. Brooks' comet of 1904 was observable for more than twelve months, Comets 1889 and 1936 I were observed for over twenty-four months, Comet 1927 IV for four years, while Grigg's comet of 1901 was observable for only twelve days, and a few very faint comets have been observed perhaps only on a single occasion.

The word 'comet', according to the Oxford Dictionary, comes from the Greek word κομήτης, which means a 'long-haired star'—presumably a comet—and is doubtless also connected with *coma*, the Latin word for 'hair'. The name thus fancifully refers to the most obvious feature of many comets, which resemble a star embedded in a mildly luminous fog from which there appears to be carried away a long streaming trail itself faintly luminous. It is this tail that no doubt suggested to the ancients long tresses of feminine hair streaming in the wind. Naturally the brightest comets (as seen from the Earth) and

1

those with the most extensive tails are the ones that have made the strongest impressions, and so appear more typical of comets than they may really be; in fact, most comets are very faint objects detectable only with powerful telescopes. We should perhaps have said 'most observable comets', for there is little room to doubt, from the steady rate of discovery of new comets, that there must be well over a hundred thousand in the solar system at present unobservable but which may all eventually become observable when they arrive nearer the sun, and it may well be that there are in addition a comparable number that move round the sun always at such great distances that they may never become visible at all to us on Earth.

Early notions

Although the Babylonians seem to have suspected that comets moved analogously to planets, the earliest ideas as to the location of comets in space were extremely inaccurate, and the assertions of some of the most learned men among the ancients can now be recognized as little better than unverified guesswork. Anaxagoras and Democritus attributed comets to 'the combined splendour of a concourse of planets'. Even so renowned a personage as Aristotle maintained that they were some kind of exhalations, whatever that may have meant, from the Earth itself that had somehow reached the upper part of the atmosphere and there in some unaccountable way become inflamed to make themselves luminous. Absurd as this idea now seems, it nevertheless may have appeared at the time to have some observational foundation. For comets are generally brightest when nearest the sun, at which time they are of course best visible shortly after sunset or shortly before sunrise, and since the tails necessarily point away from the sun these would usually appear to be more or less upright in the atmosphere, rather like the rising flame of a torch. In justice to Seneca, however, it should perhaps be said that he did not favour this idea, but he was distinctly in the minority, and Aristotle's view seems to have been so widely accepted (perhaps in itself a suspicious feature of any conjectural hypothesis) that Ptolemy

2

for example, according to the Almagest, did not regard comets as among the heavenly bodies at all.

Though Cardan had already concluded that comets must lie far beyond the moon, the first definite demonstration of their truly celestial character was due to the famous Danish astronomer Tycho Brahe who found by careful measurements that the apparent position of the daylight comet of 1577 as seen from his observatory at Hven in the Baltic sea was indistinguishable from its direction as seen from Prague, some 400 miles or so to the south. Tycho had no difficulty in perceiving the implication of this. At such separated stations a body like the moon, for instance, though very much farther away than any part of the Earth's upper atmosphere, nevertheless shows a considerable apparent angular difference of position (at times as much as five minutes of arc relative to the background stars, which are so distant as to show negligible parallax shift, as the effect is termed). From this it could be inferred with certainty that the comet was far more distant than the moon, and hence that comets were truly celestial objects.

There yet remained the problem of deciding how the comets moved with reference to the sun, for move they certainly did, but their apparent motion as seen from the Earth in most cases little resembled that of the planets, the elongated cometary paths lying partly inside and partly outside the Earth's orbit producing very mysterious differences. Tycho himself suggested that this particular comet of 1577 might move in a circle somewhere outside the orbit of Venus, but Kepler on the other hand was of the opinion that all comets move in straight lines. It had not yet come to be recognized that comets could return—identity of comets involved in widely separated apparitions rests on comparison of orbits—so it did not occur to Kepler to extend his laws of planetary motion to comets. The suggestion that comets might describe more or less parabolic orbits was put forward quite conjecturally by the German astronomer Hevelius, who attributed the necessary departures from rectilinear motion to resistance by the ether, and the hypothesis was shown to hold, at any rate for the comet of 1681 in particular, by Doerfel, one of his pupils. The theory of gravitation was soon afterwards announced by Newton, and by its means the problem of the cometary orbits was finally solved by Halley, who showed from a discussion of twenty-four comets observed between 1337 and 1698 that within the limits of accuracy of the measurements all these appeared to move in strict accordance with the law of gravitation. Apart from certain slight exceptions (to be discussed later) that have since been established, there is no reason to doubt the general applicability of gravitational theory to cometary orbits, and nowadays all orbits are computed on this basis without the smallest doubt as to its validity for such purposes.

Where the causes of cometary tails are concerned, the early ideas were almost of necessity purely conjectural. Tycho and other contemporaries believed they were mere optical appearances, free of any material nature, formed somehow by the passage of the sun's light through the comet itself. Hooke supposed that the impulsion of solar rays on the comet drove off imponderable material not subject to attraction by the sun but to repulsion, thereby anticipating the general nature of the correct explanation by means of light pressure. Electric and magnetic actions were suggested by Bessel and Olbers, while the combined effects of gravitation and a hypothetical repulsion were examined by Roche on the assumption that a comet consisted of a homogeneous gaseous atmosphere retained by a gravitating nucleus.

Possible forms of orbits

If the attractive forces of the planets are ignored in comparison with the dominating influence of the sun, the path of any small body moving near the sun must have the form of a fixed 'conic section' (that is, one of the various curves in which the surface of a circular cone is cut by a plane) and moreover the sun must always lie at its focus. Although not fundamentally different analytically, the types of curve so obtained fall into three kinds, namely ellipses, parabolas, and hyperbolas, and there are important dynamical distinctions to be made

4

between these. The ellipses are closed curves of finite extent, so that a body describing an ellipse about the sun must always remain in its neighbourhood (see Fig. 1). The parabolas are, as it were, closed at one end but just extend to infinity at the other, and a particle moving about the sun in a parabola has

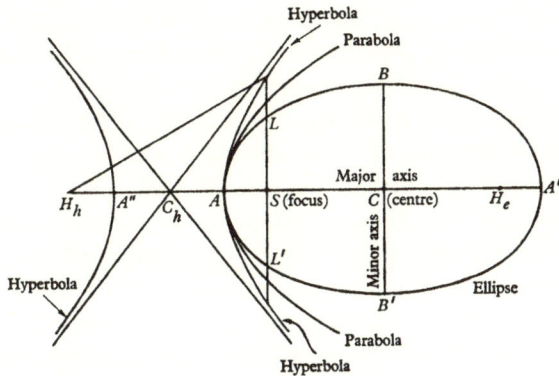

Fig. 1. Diagram of the three possible types of orbit, ellipse, parabola, and hyperbola.

$AA' =$ major axis of the ellipse, S and H_e are its foci. $SH_e/AA' =$ eccentricity (about 3/4 for ellipse shown). $BB' =$ minor axis, $LSL' =$ latus rectum.

$AA'' =$ major axis of the hyperbola, S and H_h are its foci. $SH_h/AA'' =$ eccentricity (about 7/4 for hyperbola shown).

Each of the three paths is a possible orbit under an attraction *towards* S (the sun).

Under a repulsive force from the sun S is the outer focus and the outer branch of the hyperbola is the only possible orbit.

a, e and T (the instant at which the comet passes through A) are the elements determining the path and time within the orbital plane.

just sufficient energy to escape to infinity, though it would arrive there with zero speed and take an infinite time to do so. The hyperbolas may be said to extend beyond infinity (in a complete hyperbola a disconnected branch having precisely the same form lies elsewhere in the plane), and a body moving in a hyperbolic path can escape completely from the sun and retain a finite velocity. Conversely, a particle moving towards the sun with finite velocity when at great distance will eventually sweep round it in a hyperbolic orbit, while a particle falling from rest at very great distance will have a parabolic orbit. The least decrease or increase of speed of a parabolic comet renders its orbit an ellipse or a hyperbola respectively. It is extremely important to realize, however, that the type of orbit of a body within the solar system, while possibly affording some indication of its earlier history when it can be properly interpreted, by no means provides any immediate guide in so far as the origin of the object is concerned.

When allowance is made for the presence of the planets, the paths immediately cease to be accurate ellipses, parabolas, or hyperbolas, and become highly complicated three-dimensional curves, because the force system now contains numerous additional contributions above the simple inverse square attraction to the sun. At any instant, however, a so-called osculating orbit can be defined for a body with given position and velocity as the path that the body would continue in if the planetary attraction suddenly ceased. These osculating orbits are of course perfect conic sections, and are found to bear close resemblance to the real path (because planetary action is usually small), but even so they gradually change and can represent the actual path only for a limited time, of less or greater extent, according to the degree of accuracy required and the particular disturbing effects involved.

It is possible also to arrange all the various shapes of orbits in a single series in which each one has a definite degree of elongation from the circular form which lies at one end of the set. This quality is termed the eccentricity of the curve and its value is usually denoted by the letter e. It increases all the way from 0 for a circle up to precisely 1 for a parabola and thence through all values greater than unity as the series of hyperbolas is described. What in fact Halley found for the several comets he investigated was that e was always quite near to 1, so that their paths were very close to parabolas, a characteristic feature of almost all cometary orbits. There are comets with orbits of moderate eccentricities, but they are comparatively few in number and, as we shall see, have acquired such orbits through

6

special causes. If we except such comets, the standard shape of cometary orbits may be regarded as always nearly parabolic.

The size of the orbit depends on the quantity $2a = AA'$ (Fig. 1) measuring its major axis, so that between them a and e settle the size and shape of the path. Now for a parabola the major axis is obviously not a suitable measure of its size, since it is infinite, and even for an ellipse approaching parabolic shape

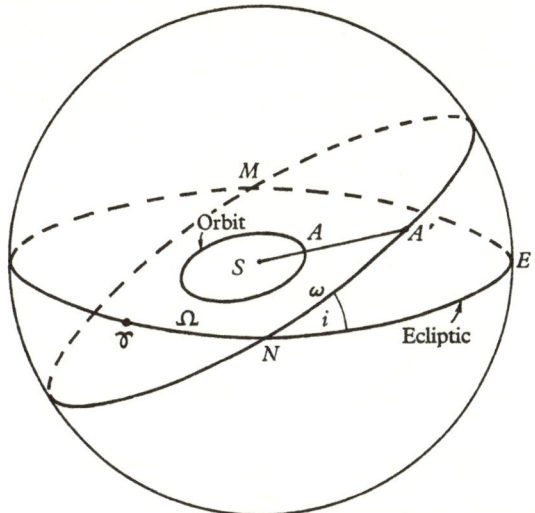

Fig. 2. The centre of the celestial sphere is S, the sun. ♈ is the vernal equinox. N is the ascending node. SA' is the direction of perihelion, A. $NA'M$ is the great circle defined by the orbital plane. $A'NE = i =$ inclination of orbit to the ecliptic. ♈$N = \Omega =$ longitude of the node N. $NA' = \omega =$ angular distance of A' from the node. Ω, ω, and i are the three elements fixing the position of the orbit in space.

it is also unsuitable because it is nearly always very large compared, say, with the size of the Earth's orbit (which provides a standard within the solar system), and is also difficult to measure accurately. It is found more convenient to use the quantity $q = a(1-e)$ which in all cases measures the least distance, SA, of the body from the sun. To fix the position of the comet completely it is necessary also to settle the time at which it passes, or will eventually pass, some specified point of its path, normally the perihelion point A. The place it reaches at any other time, and the time it takes to reach any other point, are then accurately calculable entirely by means of dynamical theory. To relate the position of the comet at any instant to that of the Earth, which is what is involved in predicting its observable future path, the precise position in space of its elliptic or parabolic orbit has also to be determined. For this purpose the plane of the Earth's motion round the sun—the ecliptic—is adopted as a reference plane; and then the inclination to this of the plane of the comet's orbit, and the angular extent to which this latter is slewed round from some standard position (the direction of the sun on 21 March, the so-called vernal equinox point) are measured, and fix the orbit completely (Fig. 2). These quantities, six in all, termed the elements of its orbit, can be tabulated for each comet, and as already mentioned, a knowledge of them suffices for the position of the comet at any future date to be calculated. The following are a few instances of typical elements of cometary orbits.

TABLE I. ORBITAL ELEMENTS OF COMETS

	1942 VII P/Oterma	1947 i P/Encke	1922 II Baade (hyperbolic)	Halley
T (perihelion passage)	1942 Aug. 21·8085	1947 Nov. 26·3295	1922 Oct.26·03015	1910 April 19·65
$q = a(1-e)$	3·389625	0·341130	2·25877	0·587200
e	0·144425	0·846197	1·000865	0·967281
Period (years)	7·89	3·284	—	76·03
ω	354°·8058	185°·1985	118°·3072	111°·6989
Ω	155°·1708	334°·7223	220°·4825	57°·27
i (Equinox 1950)	3°·9899	12°·3505	51°·4639	162°·2117

ω = angular distance of perihelion from node,
Ω = longitude of node ΥN.

Determination of orbits

When a comet is first discovered, the solution of the converse problem is the computer's initial task, and it is necessary to determine its six elements from a limited number of more or less accurate angular observations of the comet at different instants, known with high precision, and the wider spaced these can be the better for the final accuracy and reliability of the results. When a comet is newly discovered wide spacing is obviously not possible at first, since only observations covering a few days' motion will be available. For this reason a provisional orbit of only moderate accuracy is usually calculated to begin with, simply to help keep track of the comet for the purpose of securing more observations; the computation of the final definitive orbit is not made until practically all useful observations are available, and these may cover an interval of a year or even far longer. It must be remembered, however, that all computed orbits are necessarily based on observations over a limited time, and as such can represent only an approximation, even if the observations themselves were perfect, to the temporary osculating orbit—that is to the accurate conic section that the comet would describe (if it were a point mass) were all planetary disturbances suddenly to cease. Residuals between past observations and the computed orbit of anything up to a few minutes of arc are known for some comets, and where predictions of subsequent returns are concerned errors measured in days are by no means unknown especially if planetary influences have meanwhile been large on the comet. Up to the present time the orbits of something approaching 1000 comets have been investigated in this way; the results provide sufficient material to enable a completely adequate general idea of the distribution of the orbits in space to be formed.

From this evidence it appears that about three-quarters of all observed comets move in approximately parabolic orbits ($e = 1$) while about one-quarter move in paths that are definitely elliptic ($e < 1$) though the majority of these nevertheless have high eccentricities. The remaining few per cent have slightly hyperbolic orbits (that is, e just greater than 1) in so far as can be ascertained from their observable motion near the sun. But there is good reason to suppose that the actual complete paths even of these comets do not in reality extend to infinity at all but, while certainly receding to great distance, remain

9

always within the sun's influence. The perihelion distances (q) of known comets range from values just exceeding the sun's radius up to values greater than 4 a.u., the largest value to date being 5·50 a.u. for 1925 II, but comets with large q are inevitably very faint objects likely to be unobserved. All known comets with very small q (of the order of a few solar radii) are of long period.

The reasons why in some cases the path is found to be hyperbolic arise first from the extremely limited arc over which observations are sometimes only possible, and second from the additional attractive effects of the planets which may accelerate a comet that would otherwise be moving in elliptic motion into an apparently hyperbolic path. A few such orbits, apparently hyperbolic near the sun, have been extended further outwards from the observable part and backwards in time, by calculations making due allowance for the influence of the planets (Jupiter is usually the main perturbing agency), and in every case it has been found that the comet has in fact come in from a finite distance, and is therefore to be regarded as a reappearance of a permanent member of the solar system as far as its orbital motion is concerned. According to investigations by Strömgren of about twenty orbits, which near the sun were apparently hyperbolic, at great distances the speeds were always less than the velocity of escape, so that the orbits were on the whole definitely elliptical (see p. 105). Thus it can be concluded beyond any question from all this in regard to the origin of comets, that they do not simply enter the solar system from outside under free gravitational motion.

Apart from such dynamical effects of the planets, it is found that near perihelion, the point of the path closest to the sun, a given set of positions measured with even fairly high accuracy may turn out to be equally well representable by any one of a range of curves differing slightly in eccentricity. This feature of cometary orbits is illustrated in Fig. 1 which shows three different curves, an ellipse, a parabola, and a hyperbola, for each of which the arc near A is practically the same. The calculated orbit of a comet whose position happened to be capable of being measured only at points of this arc, and then always with more or less error of course (for comets being diffuse objects can seldom if ever be observed with anything approaching the same accuracy as a star or planet), could obviously be equally well represented by any one of the curves, or by any intermediate one (not drawn in the figure) and the shorter the available arc the more does this uncertainty apply. Indeed, in such cases to avoid this very difficulty it is usual to adopt the parabolic path $e = 1$ from the outset and then make the best fit possible for the five remaining disposable elements. This is why in any list of comets a fair proportion are always to be found with eccentricity precisely unity, whereas a full determination, on the known data, might conceivably lead in some cases to a slightly hyperbolic orbit. Amongst well-determined orbits strongly hyperbolic paths (near the sun) are quite unknown, the eccentricity exceeding unity in the third place of decimals at most. If earlier computations are included the greatest value of e found appears to have been 1·0525 by Hartwig for Comet 1852 II, while of more recent comets a value of 1·013 was obtained for 1886 III, though in this case the orbit was not calculable with high accuracy as the comet was observable for only about a month.

It follows from this discussion that if a path can actually be finite and yet appear to be hyperbolic as judged by its part near the sun, then the size of the orbit as measured by its major axis $2a$ must in such cases be subject to very great uncertainty. Now the time of revolution of a comet in its orbit (if undisturbed) must obey Kepler's third law—that the square of the period is proportional to the cube of the major axis—so it follows that the period of a comet when calculated from a necessarily limited arc is also only weakly determined. For instance, in the case of Donati's comet, although observable for 275 days, different computers estimated the period as 1879 years, 2040 years, and 2138 years. Observations of Coggia's comet of 1874 led to a simple elliptic period of nearly 14,000 years, but detailed calculations by Fayet taking into account planetary action on the part of the orbit in which it first approached the sun gave the real time for that particular complete circuit of the sun of only about 5000 years. Then again, for the comet of 1769 periods of 450 years, 1230 years, and 2090 years were found by Euler, Lexell and Bessel, while an even more extreme instance is provided by the comet of 1680, which according to Encke's computations, based on more accurate data, has period not 170 years as found by Euler, nor 575 years by Halley, nor 5864 years by Pingre, but in fact 8814 years!

It is also important to realize that for such comets the whole idea of a permanent regular period of motion will usually be quite inapplicable, since changing planetary action from one revolution to the next may change the time taken by a substantial amount that becomes greater the longer the period of the comet. Even for a comet of fairly short period like Halley's, which records show was observed at every return back to the year 240 B.C. (and with less certainty to 467 B.C.), the individual times of revolution have deviated by more than two years on either side of the mean period, which has been about seventy-seven years when averaged over the past 2000 years, because of differing planetary attractions during successive revolutions. The true situation is that every comet moves under the combined influence of the sun and all the planets together, and therefore that its path is not ever strictly a conic section but a highly complicated non-periodic path deviating all the time to a less or greater extent from a fixed ellipse, parabola, or hyperbola, as the case may be. For comets of moderate period the deviations are so small that, at any rate for one or two revolutions, it may be permissible to represent the path by a simple ellipse, for example, but this can seldom, if indeed ever, remain permanently an adequate approximation to the path.

Large changes of orbits

It may even happen on rare occasions that a comet passes so near to a planet that it is deflected into a new path round the sun entirely differently shaped and situated from its former path. This is illustrated in Fig. 3, where the original orbit and the subsequent orbit are shown as ellipses, as they would probably be with high accuracy for a long period before and after the encounter, though never of course with perfect accuracy because of the ever-present planetary effects. Instances are on record of this process actually having been seen to occur. For example, Brooks' comet (1889 V) passed for over two days within the satellite system of Jupiter and meanwhile described 313° of longitude round the planet, at a

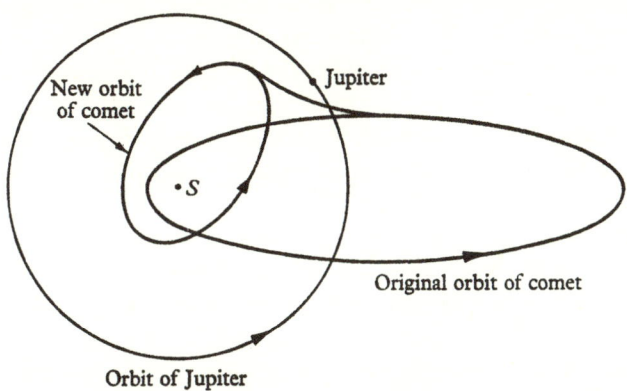

Fig. 3. Diagram showing how a comet moving in an orbit that passes near that of Jupiter may be deflected into a new elliptic orbit entirely differently situated. (Deflexion into a truly hyperbolic orbit is also possible.)

least distance of 2·28 radii of the planet, incidentally without producing detectable perturbations on any of the satellites, but as a result the period of the comet's orbit was changed from about twenty-nine years to about seven years. Again, Lexell's comet, discovered in 1770, underwent a close encounter with Jupiter in 1779 and moved into a fresh orbit in which the comet has since become unobservable. At all events it was lost afterwards and no comet since discovered has proved certain of identity with it, though some claims appear to have been made for Comet 1895 II. An even more remarkable instance was provided by Comet Wolf which underwent a close approach to Jupiter in 1875 and again in 1922. By a strange chance the second encounter almost exactly reversed the effects of the first to send it back into practically its original orbit. Before 1875 and after 1922 it moved in an orbit having perihelion distance of about 2·5 a.u. whereas the temporary orbit from 1875 to 1922 had perihelion distance of only about 1·5 a.u. Its aphelion distance on the other hand, remained more or less unchanged throughout at about 5·5 a.u., a value quite close to the radius of Jupiter's orbit. It could even happen through such an encounter that a comet was thrown into a sufficiently strongly hyperbolic orbit for it to escape from the sun's field and be entirely lost from the solar system.

A sufficiently close approach to a small planet affords the best hope of a direct determination of the mass of a comet being made, but it seems probable that only for a very large comet is there much likelihood of any detectable effect on the planet's path. Conversely, however, the deflection of an accurately known cometary orbit by a planet could provide evidence for the determination of the planetary mass. Such an event occurred in 1835 when Encke's comet passed fairly close to Mercury with results that showed by comparison with observation at later returns that the hitherto accepted value of the planet's mass was far too great. Because of its smallness and proximity to the sun, the mass of Mercury is difficult to determine with high accuracy, but it is possible that improved accuracy will one day be attained if a really close encounter by a suitable comet should take place.

Contrast between planetary orbits and cometary orbits

Returning now to the general nature of cometary orbits, the most noticeable thing is that by far the majority possess features that are almost the direct opposite of planetary orbits. While the latter lie nearly in the same plane and are almost circular, the planes of the comets' orbits, apart from those of short-period which are not typical, are distributed almost randomly, and the shapes are always nearly parabolic. According to some writers there is evidence of clustering of the directions of the major axes in certain regions of the celestial sphere. Fig. 4 shows a diagram constructed by Eddington from data compiled

14

by W. H. Pickering of the directions of the aphelia of some three hundred long-period comets. This definitely shows a certain amount of close grouping despite the fact that past perturbations must have tended to efface any original peculiarities of distribution. Also there are fairly large areas devoid of any aphelion points, but selection effects may have contributed to this. A further study based on all available orbits

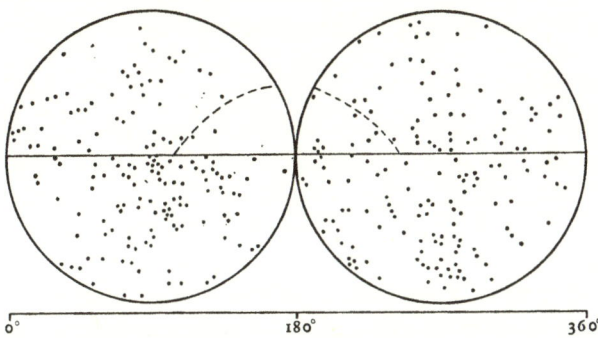

0°   180°   360°

Fig. 4. Distribution of aphelia of 317 comets (Eddington). The diagram represents an 'equal area' projection of the eastern and western hemispheres of the celestial sphere; the line of centres of the two circles corresponds to the position of the ecliptic. Each dot represents a cometary aphelion. The area between the ecliptic and the broken curve is Hoek's zone (p. 21).

would be desirable, for if such clustering could be established for the present orbits, it would strongly suggest a greater degree of clustering at earlier stages.

Where the planets are concerned their orbital inclinations to the ecliptic are all very small (apart perhaps from Pluto, 17°) and they share the same *direct* motion round the sun. But the inclinations of the cometary orbits have values lying all the way from 0° to 180°, so that some comets have direct and others retrograde motion. Many comets have orbits of such high inclination (near 90°) that the word 'direct' or 'retrograde' is hardly applicable to their motions, but ignoring this consideration it is found that where the long-period comets are concerned there is just a slight preponderance of retrograde motions.

Two comets are known that move in nearly circular orbits: Comet Schwassmann-Wachmann (1925 II) for which $e = 0.136$, $i = 9.5°$; and Comet Oterma (1942 VII) for which $e = 0.144$, $i = 4°$. They are capable of being observed all round their orbits, but like superior planets are usually most favourably placed each year when near opposition. For this reason they are sometimes called 'annual' comets. But resemblance to planetary orbits is highly exceptional, and these two almost circular paths must be regarded as curiosities among cometary orbits.

Short-period comets

If comets are classified according to periods, they may be divided into two fairly distinct groups, the short-period group and the long-period group. In the former class there are over seventy comets with periods less than 100 years, and of these rather more than forty have periods between five and seven-and-a-half years. Twenty-four of these have been observed at two or more returns. There are strong reasons for believing that all members of this short-period group are exceptional and represent examples of comets diverted by the attraction of Jupiter from former almost parabolic orbits into their present elliptic orbits, which are mostly of only moderate eccentricity. Also, the planes of their present orbits are inclined at moderately small angles to that of Jupiter's orbit, the average absolute inclination is about 13°, only three exceeding 30°, the greatest being 54° for Tuttle's comet (1871 III), and with the exception of Halley's comet all have the same general direction of motion round the sun as the planet, both of which features were shown by H. A. Newton to be conditions strongly suggestive for the hypothesis of capture by Jupiter. Although the original encounters bringing about the main deflexion must in most cases have been so long ago that considerable further changes in the orbits have since resulted, the present orbits in nearly all cases are still such that strong perturbation by Jupiter could occur if a favourable conjunction came about.

Many comets have been discovered moving in short-period orbits but not found again at any subsequent return, while

16

several have been missed after a number of observed returns. For example, Brorsen's comet of 1846 (period 5·5 years) was not seen again after 1879, and Holmes' comet of 1892 (period 7 years) was not found in 1919 or 1928. Other comets, such as Encke's in 1944, have occasionally been missed through unfavourable circumstances but rediscovered at a later return.

Because of their close dynamical association with Jupiter this group of comets is often referred to as 'Jupiter's family of comets', and there is no doubt whatever as to its reality. On the other hand it has been rather hastily conjectured by a number of writers that corresponding families exist associated in a similar way with the remaining great planets, Saturn, Uranus, and Neptune. While this may eventually be established should observations of fainter comets of greater perihelion distances become available, H. N. Russell has shown that there is no reliable evidence for such families in the whole of the modern available orbital data, none of the present alleged members ever coming within anything like small enough distance of its assigned 'parent' planet. Some astronomers, notably Forbes, Geuillot, and Lowell (who appears to have been obsessed with the possibility of the existence of trans-Neptunian planets), have gone even beyond this by reversing the already highly doubtful argument to use it to infer the existence of (unknown) remoter planets still, on the basis of a few comets apparently having periods grouped near 335 years and 1000 years, values in each case so long that it would be particularly difficult to settle them accurately.

Comets of moderate and long periods

The comets of longer periods, lying between say about 100 years and 1000 years, of which there are about forty known, show no such close connexion with Jupiter's orbit, none of them coming sufficiently near for *strong* perturbations ever to be likely. Their comparatively short periods and elliptic orbits make it clear that dynamically they are permanent members of the solar system.

2

But by far the majority of comets possess much longer orbital periods than this, and according to Crommelin the average is about 40,000 years. Such an estimate can only be a rough order of magnitude average, since their periods, which would vary a great deal from one revolution to the next, are obviously not obtainable from successive returns, and the orbits of such comets in their single observed sweep near the sun will usually perforce be computed on the basis of being parabolic, as explained above. Comets of such long periods can be of only temporary interest observationally, disappearing from us for ever after their one appearance, but nevertheless by far the majority of all observed comets come into this category.

The average number of comets discovered each year is about six—the high record is thirteen in 1932 and again in 1947—but of these only three or four are really new, the others, as is established by unmistakably close agreement of all the orbital elements, being returns of earlier discovered comets of moderate periods. This discovery rate means that at least three hundred long-period comets come to perihelion each century, and if we adopt 30,000 to 40,000 years as the average period, we arrive at the amazing but inescapable conclusion that there must be at least 100,000 comets in the solar system with perihelion distances sufficiently small for them to become eventually observable. It can scarcely be doubted that there are many more with perihelion distances too great for the comets to become observable with present equipment so long as they remain in their existing paths.

The sun-grazing comets and other comet groups

In addition to the foregoing classification there are several instances of a number of distinct comets having almost identical paths and following each other along them spaced out at intervals that in time range from a few months in some cases to hundreds of years in others. Perhaps the most famous of all such groups consists of the very large comets observed in the years 1668, 1843 I, 1880 I, 1882 III, and 1887 I, which were also highly remarkable for their extremely small perihelion

18

distances—each of them passed within less than a solar radius of the sun's surface, which means a value of q of about 1/200 of an astronomical unit, indeed those of 1880 and 1887 approached within 90,000 miles of the sun's surface, less than one-fifth of the sun's radius. The orbits of all these comets are so elongated that, apart from the portions very near the sun, the paths are almost straight lines, but nevertheless the first two were found to be measurably elliptic, though the periods could be only extremely poorly determined and estimates ranged from about 400 to 1000 years. Both these comets were bright and presumably very large, while those of 1880 and 1887 were faint in comparison, but the one of 1882 was the largest of the five. According to Kreutz the bright comet of 1702 is also to be counted a member of the same group.

There seems little doubt that groups of this kind may in some cases represent the component parts of a larger primitive comet, but we will leave aside the discussion of the cause of the disruption till later. However, it is of interest to mention here the strong observational support for a disruption process afforded by the actual observed division of the 1882 comet of this group, which soon after passing perihelion was seen to split into at least four parts which gradually separated from each other along much the same path but with slightly differing periods that have been calculated to be about 670, 770, 880, and 960 years. So this great comet of 1882 may be expected to reappear as four smaller comets, though still fairly large as comets go, separated by intervals of some hundred years but moving in much the same orbit as all the other members of the group unless differing perturbations of individual members have meanwhile interfered.

Numerous other similar groups, but of distinctly larger perihelion distances, with more or less identity of orbits amongst its members are also known, for example the comets of 1742 and Comet 1907 II have orbits almost the same; that of 1812 and 1884 I go together; and so do 1884 III and 1892 V. In each case the separation in time along the orbit is quite different from the period, so that the comets themselves are certainly distinct.

Distribution of cometary aphelia

The possible existence of cometary groups appears first to have been recognized in 1866 by Hoek. From a study of the distribution on the celestial sphere of the aphelion points Hoek discovered at least seven different groups of comets, each containing at least four comets and one as many as twelve, having practically the same aphelion points. Thus for one group the celestial coordinates of the aphelia were as follows:

Comet	Longitude	Latitude
1672	279·4	−69·4
1677	286·4	−75·7
1683	290·8	−83·0
1860 III	303·1	−73·2
1863 I	313·2	−73·9
1863 VI	313·9	−76·4

These values alone are strongly suggestive of grouping, but in fact the actual positions on the celestial sphere are even more closely associated than the longitude figures suggest owing to the high latitude. The great circle differences between the points amount to little more than 3°. But in addition to this Hoek found that the points of intersection of the orbits (on the celestial sphere) for these six comets lie within a small circle of radius only 1° centred at the point (319°, −78°).

The possibility of a common origin for at least some of these comets in a region at great distance from the sun receives cogent support from further calculations by Hoek of the several distances of these comets from the sun at past epochs. Thus, for the last three comets of the group he finds the following values (in astronomical units) at the earlier dates shown.

Date	Distance from the Sun (a.u.)		
	1860 III	1863 I	1863 VI
1858	10·0	15·9	17·4
1785	100·0	101·8	102·1
757	600·0	600·4	600·2

The final differences of distance are so small that Hoek concluded that each group of comets must have originated at practically the same time from a common region with slight velocity differences amounting at most to about two feet per second. The calculations bring out a characteristic feature of nearly parabolic motion, namely that extremely small differences of speed at great distance can lead to great differences in period.

From the distribution of the aphelia of 190 comets for which data were then available, although finding the representative points widely and fairly uniformly scattered over the celestial sphere, Hoek nevertheless noted not only these numerous groups of closely associated points, but also considerable areas devoid of any aphelia, especially a gap just north of the ecliptic extending from longitude about 95° to 243° and rising to a latitude of about 32° at longitude 169°.

In the diagram constructed by Eddington (1913) showing the positions on the celestial sphere of the aphelion points of 317 comets (reproduced in Fig. 4) the upper boundary of the region drawn attention to by Hoek is indicated by a broken line. This figure also shows other equally large areas almost free of aphelion points, notably that surrounding the point of longitude 0° and latitude about 55° north, and it also shows numerous groups of points lying close together. It is difficult to judge by simple inspection whether the apparent peculiarities of the distribution are of such a nature as cannot be attributed purely to chance, and there is need for a careful statistical study of this question on the basis of the greater amount of data now available. The theory of the origin of comets arrived at later in this volume would certainly not only be consistent with the existence of clusterings of aphelia but in the first stages immediately after formation would definitely require this. The ever-present effect of planetary perturbations must in the course of time tend to displace the points and lead to a more uniform distribution, though it may well be that for very long-period comets there has not been sufficient time for the initial clusterings to have been much dispersed.

Speeds of comets in their orbits

If for the moment we exclude certain minor effects of retardation that have been found for a few well-observed comets, it can safely be said that, within the limits of observational error

21

COMETS AND THEIR ORIGIN

in measuring their positions, the orbits of comets are adequately accounted for on the basis of the law of gravitation. The speeds with which they describe the various parts of their orbits can therefore be quite easily calculated once the orbital elements are known. According to Kepler's second law, the line drawn from the centre of the sun to the comet always sweeps out equal areas in equal time intervals, so that a comet must pass much more quickly through perihelion than through aphelion. Thus, the comet of 1843 in *one day* near perihelion described 292° of its circuit about the sun leaving only 68° to be described in the remaining hundreds of years of its total period, while the Comet 1887 I described the perihelion side of its orbit, 180° of longitude round the sun, in less than three hours, as must any sun-grazing comet.

If M is the mass of the sun ($1 \cdot 992 \times 10^{33}$ g.), and G the constant of gravitation ($6 \cdot 67 \times 10^{-8}$ c.g.s. units), the velocity v at a general point distant r from the sun is given by the simple formula

$$v^2 = GM(2/r - 1/a),$$

and at perihelion where $r = q = a(1-e)$, the velocity, v_p say, is therefore given by

$$v_p^2 = GM(1+e)/q.$$

The velocity at aphelion, v_a say, at distance $a(1+e)$ from the sun is then simply related to v_p by

$$v_a = v_p \frac{1-e}{1+e}.$$

Since e is close to 1 for most comets, the formula again shows that they move much more slowly at great distances than at small distances. Some actual numerical examples can readily be given.

(i) *Encke's comet.* Period = $3 \cdot 3$ years, $a = 2 \cdot 219$ a.u., $e = 0 \cdot 8458$. We find at once $v_p = 69 \cdot 2$ km./sec., $v_a = 5 \cdot 8$ km./sec., $v_p/v_a = 12$.

(ii) *Halley's comet.* Period = about 77 years, $a = 17 \cdot 947$ a.u., $e = 0 \cdot 967281$. Here $v_p = 54 \cdot 5$ km./sec., $v_a = 0 \cdot 91$ km./sec., $v_p/v_a = 60$.

(iii) For long-period comets the outer part of the orbit is so weakly determined that accurate values of the velocity there cannot be computed. However, Donati's comet when at perihelion moved at about 30 miles a second, and when at aphelion at less than one-seventh of a mile a second, a ratio of 200 to 1. For the sun-grazing comets the ratio must be far larger still, as indeed Humboldt found for the comet of 1680 which had perihelion velocity of nearly 250 miles per second and aphelion velocity of about 10 feet per second, a ratio of about 10^5 to 1.

These figures emphasize the far greater proportion of their time that comets spend on the aphelion side of their orbits, and as it is usually only when describing the perihelion side that a comet is observable it follows that the majority of comets are most of the time invisible objects. The few that become observable each year out of the large number known clearly demonstrates the truth of this.

The acceleration of Encke's comet

No account of the orbital motions of comets would be complete without mention of the remarkable deviation from a strict Newtonian orbit discovered by Encke for the faint short period comet that bears his name, although not discovered by him. If we exclude 1949g which had an estimated period of 2·3 years but became lost before adequate observations were made, Encke's comet, sometimes called the Mercury of comets, has the shortest known period (3·3 years), and has been observed since its rediscovery in 1818 by Pons at every one of over forty successive returns till that of 1944 when, amongst other unfavourable circumstances, it was situated too far south for most observatories to search for it. What Encke found was that after making every allowance for the perturbing influence of the planets there yet remained outstanding a progressive shortening of its period. During about a century the period of 3·3 years was found to have decreased in all by as much as about 2·5 days. Moreover the rate of decrease of the period was found not to be quite regular but itself subject to curious changes, indicating probably some varying or intermittent cause, or a combination of varying causes. Any such decrease of period must be accompanied by a decrease in the size of the orbit (strictly, in the major axis), and Backlund showed that at the present rate of diminution the path must have extended out to Jupiter's orbit only a few thousand years ago. This suggests that at such a time the comet was probably deflected by Jupiter into a short-period orbit that since then has gradually diminished and so prevented subsequent close encounters with the planet.

There is some evidence of similar accelerations in a number of other comets, notably Comet 1939 VII which has a period of 6·95 years, and Comet Brooks (2) 1946e, while according to Maubant Comet 1908d showed a small *negative* acceleration—that is a retardation of longitude—a phenomenon also alleged to have been detected in the motions of Comet Tempel and Comet Brorsen.

Discovery and designation of comets

Comets have been discovered in several ways ranging from planned telescopic searches using a specially low-power eye-piece with a large field, itself called a 'comet-seeker', designed for sweeping carefully over a selected area of the sky, as the most direct method, to chance discovery from time to time on photographic plates taken probably with some quite different object in view. Reference back to such plates often reveals earlier positions of comets when these have later been discovered by some other means. It has also happened, notably in 1882, 1893, and 1948, that unseen comets close to the sun have been more or less accidentally found during solar eclipses, suggesting that many comets must come to perihelion close to the sun yet escape observation altogether. But the most remarkable of all chance discoveries of a comet deserves mention to itself. Apparently in 1896, while Perrine at the Lick Observatory was regularly making observations of a comet he had himself discovered, a telegram originating from Kiel stating the present place of this comet became seriously garbled in transmission and gave an entirely wrong position more than 2° away from the true one. Unaware of this, however, Perrine made observations in the erroneous place and, against every probability, found an entirely new comet there! Comets could be said to be a fairly numerous class of object, but the sky is very large, and this discovery represents perhaps one of the most amazing coincidences of all time. Curiously enough, during a search for Zona's comet (1890 IV) a second comet (1890 VII) had been found close to the expected place by Spitaler. In commenting on this latter discovery Barnard referred to the remarkable coincidence of two unconnected comets being observable at the same time within little more than a degree of each other, 'a thing which had never happened before and *never likely to happen again*'. Perrine's experience took place only five years later!

The greatest number of comets that have come under observation in any one year is twenty-two in the year 1947; these consisted of eight already observed in 1946 and carried over into 1947, seven new long-period comets, one new short-period comet, one cometary object regarded as doubtful, and five returns of comets identifiable with earlier known ones. In 1948 the number of *new* comets reached the record total of ten. On the other hand, in 1934 no new comet was found though seven comets were under observation, while in 1938 only one comet, and that not a new one, came under observation during the whole year. The discovery of faint comets by photographic means, however, is likely to increase, as also are discoveries of southern comets with the establishment of more southern hemisphere observatories; many such comets must hitherto have escaped detection. As returns of earlier known comets are included in the annual reckoning of discoveries, it seems highly probable that fresh records will be achieved before long.

As for nomenclature of comets, it has long been the custom to assign to them the names of their recorded discoverers, occasionally with suffixed number if it is not the first of that name (thus Tempel$_2$ 1873), or in a few cases, failing that of the discoverer, the name of anyone who may have come to be specially associated with the comet through long or important research concerning it. Halley's comet is necessarily in this latter class, its first discovery doubtless having been prehistoric, but Encke's comet was first discovered by Mechain, of Paris, in 1786. When a comet is rediscovered at a later return, or discovered for the first time more or less simultaneously by more than one observer, it may then be given two names, or even three. But if this procedure were continued it would make the name eventually very cumbrous, and by common agreement the process now stops at three joint names. Thus, for example, the Pons-Brooks comet was first discovered by Pons and rediscovered on its return in 1883 by Brooks. Pons, by the way, who commenced his connexion with astronomy in the capacity of concierge at the Marseilles Observatory, appears to have been the most successful comet searcher of all in finding at least thirty-seven, if not several more. His closest rivals appear to have been Brooks with twenty, Barnard with nineteen, followed by Perrine with thirteen and Swift with eleven.* An idea of their patience can be formed from Denning's estimate—he discovered five—that each comet found corresponded on the average to about 120 hours searching. Again, the remarkable comet found in 1925 by Schwassmann and Wachmann, whose almost circular orbit lies entirely between those of Jupiter and Saturn, bears both their names. Among three-named comets impressive but rather unmanageable appellations such as the following have come about through the continuation of this practice: Comet Jurlof-Achmarof-Hassel; Comet du Toit-Neujmin-Delporte; Comet Whipple-Fedke-Tevzadze; Comet van Gent-Peltier-Daimaca—to mention but a few. At the 1948 I.A.U. meeting the following self-explanatory resolution was unanimously agreed to by comet workers: '*That the periodic comet Pons-Coggia-Winnecke-Forbes should in future be called Comet Crommelin*'. When more than one comet is named after the same discoverer or discoverers, the number (in

* These figures are given by Crommelin and Proctor (*Comets*, pp. 155–6, 1937) but in *M.N.* **64**, 840, 1904, Brooks himself announces his twenty-fourth cometary discovery.

order of discovery) is nowadays usually added in brackets to indicate this instead of a suffix; thus Comets 1925 II, 1929 I, and 1930 VI were all discovered by Schwassmann and Wachmann, and are named S-W (1), S-W (2), and S-W (3) respectively.

Obviously with the ever increasing number of comets this ancient practice of using the discoverer's name, prompted originally no doubt by the best intentions, can lead only to greater and greater confusion, as it affords no means of classifying comets. Accordingly it has been necessary to supplement it by a notation directly related to the orbit and time of appearance that can be applied as unambiguously as possible to every comet. To this end, as each comet is discovered it is labelled in the first place with the year of discovery together with a small letter of the alphabet, starting with the letter 'a', indicating the order of discovery among all the comets of that year; the letter 'P' is also sometimes added nowadays to distinguish a periodic comet. Thus 1939a, 1939b, 1939c, and so on, represent the first three comets found during 1939, while 1947i P/Encke refers to a return of Encke's comet, and such designations can be temporarily applied quite unambiguously to each and every comet whether it is new or a return of a previously known one, which are questions that cannot always be settled at once for every comet by any means, or even if the object concerned is only suspected to be a comet. Later on, when sufficient observations have accumulated for the detailed computation of the orbit to be made and the date of perihelion passage ascertained, the comet is then labelled afresh with the year of perihelion passage together with the order in that year, amongst all known comets, of actually passing through perihelion and shown by Roman numerals. Thus 1939a eventually became 1939 I, which means it was not only the first comet seen in that year but the first comet to pass its perihelion point in 1939, and this notation obviously yields more useful information than its perhaps more romantic name of Comet Kozik-Peltier. On the other hand 1939e (Comet Kopff) turned out to be 1939 II. The change from being the fifth found in 1939 to the second to pass perihelion happened simply because there is no necessity whatever for comets to come to perihelion in the same order as discovery, and in fact it frequently happens that the final year assigned is different from the year of discovery. For example, 1939 VIII was in fact not observed by Kulin till 1940 when it was 1940a, but 1940 I was originally 1939l. The greatest number of perihelion passages in any one year appears to be eleven, in 1925. This notation of adopting the year and order of perihelion passage has been found to provide a far more advantageous system of classification which, apart from the rare exceptions of comets such as Lexell's whose orbits are seriously perturbed, is of practically a permanent character.

II

PHYSICAL PROPERTIES OF COMETS

Where is the centre of a comet?

In discussing the orbits of comets it has been assumed more or less of necessity that a comet may be regarded as a simple gravitating particle in order that its motion relative to the sun may be describable by means of a point tracing out a curve. The assumption must contain some element of truth because of the degree of agreement with dynamical theory exhibited by the observed paths, but even so when we come to consider the internal structure of comets this assumption appears to be far from closely satisfied. If for the moment we pause to consider the motion round the sun of a large planet like Jupiter, it can be rigorously demonstrated as a matter of dynamics that there exists a certain point of the body—its centre of mass— whose motion round the sun is the same as if all the mass of the planet were concentrated at this point and all the external forces acted on it, exactly as if it were a particle of negligible size. The idea of centre of mass produces a remarkable simplification where the motion of the planet is concerned because Jupiter is effectively a rigid body despite its diameter of nearly 140,000 miles. The centre of mass remains fixed in it because it is rigid, and so the motion of this point adequately represents the general motion about the sun of the planet as a whole. (The rotatory motion of the planet about its centre is a secondary problem.) But where a comet is concerned there is no prior reason whatever for supposing that the line from the observer through the brightest point, or through the central point of the visible area, which varies with the means of observation, passes through the centre of mass of the comet as a whole. As will be explained later, the outline boundary of a comet is not always well defined, and the comet's position at any time is usually settled by the simple expedient of

29

guiding on its brightest part, which does in fact sometimes appear as a definite point within the head. But there is no reason for assuming that such a point bears any fixed relation to the centre of mass of the comet; indeed, despite the long tradition behind this observational practice, it is a matter yet to be demonstrated that any such relation exists, and, if so, what its dynamical and geometrical nature may be. But for the most part the general notion seems to have long persisted that an orbit based on such measures represents a genuine orbit of some point in the comet about which the material of the comet is in some way permanently clustered. However, an observer of 1926 III (Comet Ensor) felt it necessary to record that 'the comet was so large, diffuse, and irregular in shape in 1926 March and April, that there is considerable doubt which point should be taken as the centre of gravity'. It will be of special importance to our study of the problem of the structure of comets to realize that although the orbits themselves are well-understood (for the limited periods of time they have been followed) they are nevertheless only approximate because of this difficulty in deciding what the observed point really means. Outstanding residuals of the order of a minute of arc are not infrequent for cometary orbits, but would be quite intolerable for a planetary orbit, where the accuracy is within a second of arc. It would be a cardinal mistake to infer from the apparent success of the orbital theory that comets are rigid bodies even approximately like the planets and there is little doubt that it is through the uncritical assumption that they may be regarded as more or less rigid masses that progress with the theory of their structure has been so extraordinarily limited.

General appearance of a comet

The first impression of a faint comet is usually that of a very hazy object whose exact form and size are extremely difficult to determine with precision. Later it may take on a more definite form—'like a spherical mass of vapour', as Barnard described Holmes' comet—though in many comets the apparent shape is far from circular. The size seems also to be dependent on the means of observation. While the majority of comets usually have a silvery-grey appearance, a fairly definite colour, such as blue, red, or yellow, appears to be assignable to some comets at certain times. The main features associated with comets, whatever mode of observation is used, although not physically distinct, can be conveniently dealt with under four heads. Of these properties perhaps only the coma is found to be observable with every comet, though one or two comets have been reported as entirely 'asteroidal' in appearance, presumably meaning that they have very small angular size. On the other hand, one or two comets have at times possessed no apparent head, as for instance the sun-grazing comet of 1887 which nevertheless showed a tail over 50° long, while the heads of Comets 1843 I and 1880 I (also sun-grazers) were at times barely discernible.

First there is the so-called *coma*. This is the principal part of practically all comets and has the appearance of a faintly luminous cloud with an angular size sometimes of several minutes of arc, and exceptionally perhaps even a degree or more (in 1770, according to Messier, the nebulosity surrounding the nucleus of Lexell's comet had a diameter of nearly $2\frac{1}{2}°$). In shape its outline for some comets seems roughly circular, or perhaps in some instances elongated, while fan-shaped comas are often reported, but in any case it may be far from regular and have quite different outward appearances at different returns and, above all, at different parts of its orbit. Holmes' comet (1892 III), a few days after discovery, which happened when it had passed perihelion, appeared as a sharply defined circular area 5' in diameter, but within a week it had become very irregular in outline and roughly doubled in angular size, while on a photograph by Barnard it appeared to be nearly 30' diameter. It will be hardly necessary to mention that what is seen, of course, is the comet projected against the background sky, and this by itself can afford no direct indication of the three-dimensional shape of the comet. Moreover, one of the most striking properties of the coma is that it always contracts as the comet approaches its least distance from the sun. The coma sometimes appears to be a more or less continuous haze of light, though its intensity of brightness per unit angular area may vary considerably, and it usually shades off in an uncertain way to an ill-defined boundary. For comets with well-defined rounded comae, the coma and nucleus, if present, are often referred to together as the *head* of the comet.

The coma has the remarkable property that it is transparent, and even faint stars lying in the direct line of sight have been seen without the smallest perceptible or measurable decrease in their luminosities. However, it is claimed by Chambers that an 11^m star was completely eclipsed by Denning's comet 1890 VI. E. C. Pickering found in the case of Comet 1902b that it affected a certain seventh magnitude star by at most $0^{m}\cdot05$ and on the average by not more than $\pm 0^{m}\cdot02$, amounts easily attributable to other causes and suggesting that absorption of the star's light by the comet was at all times quite insensible. On the other hand Silbernagel, during an occultation by Comet 1913a, found the star to become blurred and reddish, but on emergence from the coma it was sharp and bright as before. Also, according to Esclangon, a star occulted by 1911c (Brooks) appeared to exhibit some diminution of light. On the other hand, Donati's comet is said to have passed across Arcturus without affecting the star's brilliance, and Sir John Herschel records that Biela's comet passed directly across a small group of faint stars without perceptible effects. According to Sir W. Herschel in the case of the comet of 1807 stars seen through the tail lost some lustre, and one near the head was but faintly and intermittently visible.

Perhaps even more striking as a demonstration of the transparency of most comets is the fact that during the passages of the great comet of 1882 and that of Halley's comet in 1910 between the Earth and the sun, when their orbits took them directly across the apparent disc of the sun, both these comets (the first an exceptionally large one) became completely invisible, and this despite the most careful observation with highly accurate knowledge of the positions. There are also a

PLATE I

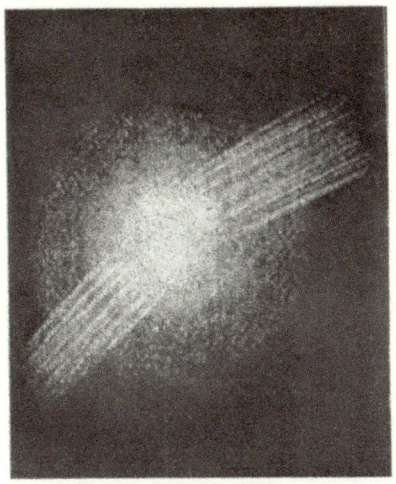

The comet of 1823

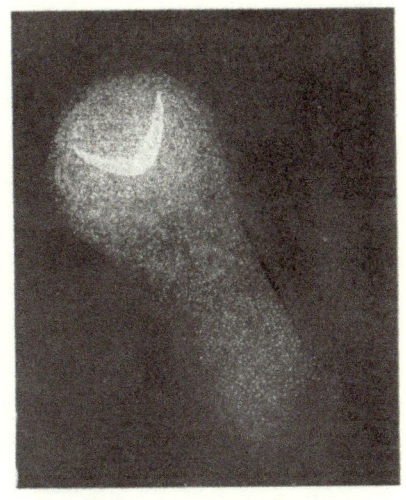

Halley's comet in 1835

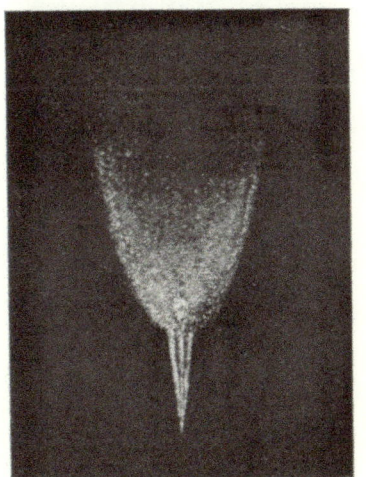

The comet of 1851

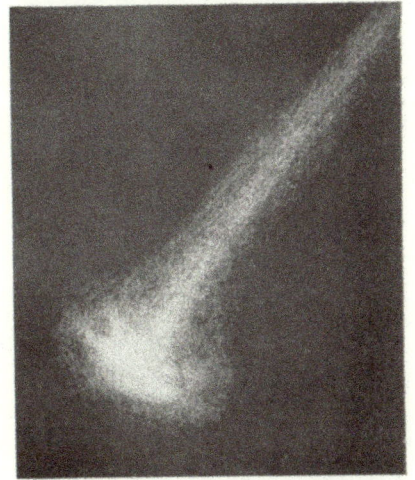

Winnecke's comet in 1868

SKETCHES SHOWING THE IRREGULAR APPARENT
SHAPES OF COMETS

PLATE II

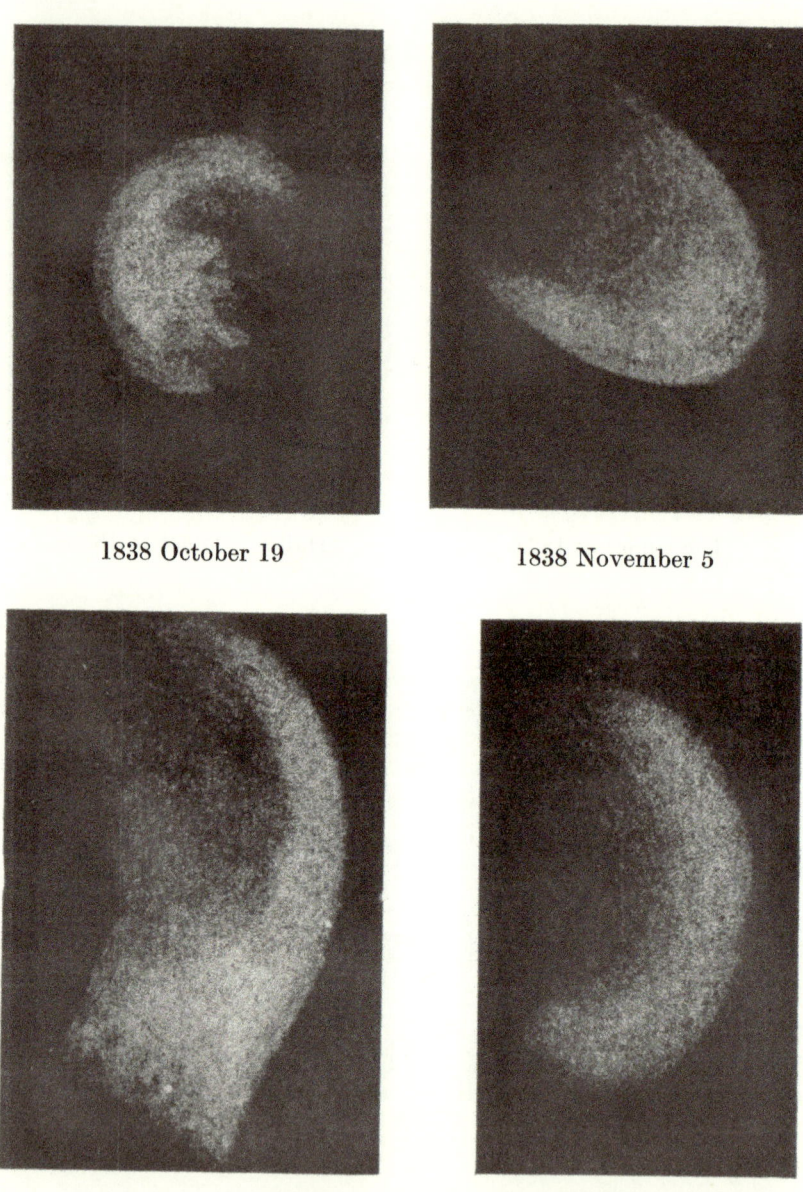

1838 October 19   1838 November 5

1838 November 12   1838 November 10

THE APPEARANCE OF ENCKE'S COMET AT DIFFERENT TIMES
DURING ITS RETURN OF 1838

number of earlier instances of comets escaping detection during transit across the sun, notably the Comet 1823.

Second, there often appears to be present within the coma or head a small bright region or point termed the *nucleus*. In some comets more than one such point may be present, but in others no nucleus whatever is to be seen. The nucleus often has the appearance of a star-like point of light condensed within the coma, but not necessarily situated near its centre. According to Arago, the nucleus is usually displaced to the sunward side of the coma. In some comets it is very conspicuous, but in others it appears as little more than a general increase of intensity of brightness at some part of the coma, but many telescopic comets show no nucleus at all. If present, the nucleus affords a convenient point to centre upon in measuring the position of the comet for the purpose of computing its orbit, though as may be emphasized again there is no *a priori* reason for supposing that it has any special dynamical significance. The possible temporary or mobile character of the nucleus relative to the comet, which itself may have no permanent or definite shape, is also suggested by the fact that the nucleus often only makes its appearance when the comet is near the sun. Barnard states that although Holmes' comet (1892 III) was at first without any trace of a nucleus, one appeared to form while he was actually observing it and thereafter remained clear and distinct. This in itself suggests that the nucleus of a comet does not represent any definite part, or solid object, permanently associated with it. Sir John Herschel came to the same conclusion from his observations of the great Comet 1861 II. Certainly no large solid body has ever been detected at the position of the nucleus, or anywhere within a comet. When Halley's comet transitted the sun, a body as small as 50 km. in diameter could hardly have escaped observation if present. Thus it may well be that the nucleus is no more than an apparent phenomenon, possibly some kind of changing concentration of small particles within the coma. In the case of Lexell's comet of 1770, Messier found that while in the course of a fortnight the diameter of the observed coma increased nearly thirty-fold from 5′ to 2° 23′, the diameter of the nucleus appeared to remain invariable at about 22′.

Third, may be mentioned the conspicuous *envelopes* or *shells* that have sometimes been seen to be emitted by the brightest comets, though only few have exhibited this phenomenon. They appear as more or less concentric illuminated sheets and sometimes strike the eye as seeming to be emitted by the comet at roughly equal intervals, which may range from a few hours, as in Morehouse's comet 1908 III, to a few days, as for the great Comet 1861 II, or even two or three weeks, as in Donati's comet of 1858. In not all comets showing this effect do the envelopes necessarily have their (apparent) centres at the nucleus. For instance, Coggia's comet (1874 III) showed both symmetrically placed and eccentrically placed envelopes and the portion to one side of the nucleus was distinctly brighter than the other, as though the comet were about to become a double comet. The motions of the expanding envelopes seem as if the comet initially ejected and repelled material from its central regions but, remarkably enough, only in the hemisphere (sometimes rather more than a hemisphere) towards the sun, and thereafter the material appears to be turned back almost directly away from the sun as though now violently repelled from it. In such cases (e.g. Donati's and Coggia's comets), the whole head of the comet appears rather like an illuminated fountain of water at first projected upwards but later turned back downwards on all sides by the force of gravity, its geometrical form often being likened to a paraboloid, though of course only the projected shape can be observed. But it is important to realize that this account, however realistic the picture may seem to observers, is no more than a description of what apparently occurs, and its correct interpretation may not be simply related to such a picture. Indeed, strong suspicions of the direct interpretation of the expanding shells as actual material motions are provided by some early work by Eddington, who found that in Morehouse's comet such motions, if real, would require the presence within the comet of extraordinarily powerful repulsive forces of a strength and kind quite unknown. Moreover, in some cases the envelopes appear stationary or even contract after formation.

Finally, but not least in importance, may be mentioned the highly sensational accompaniment of most bright comets, namely the *tail*, from which the name given to comets derives. It is the case, however, that no tails have been detected associated with certain faint comets, for example Finlay's comet (1893 III) and other short period comets; indeed, where faint comets are concerned the majority are without tails, though this may of course be simply because in most cases they are too faint to be seen. But the tails associated with large comets are often clearly visible to the unaided eye, sometimes in daylight, and the greatest lengths to which their observed portions have attained sometimes exceed a hundred million miles (nearly two hundred million miles for the great comet of 1843). On the other hand the apparent angular length of the tail in the sky depends of course on where the Earth happens to be situated, and may bear little relation to the actual length. The tail of Halley's comet in 1910 at one time extended nearly 180° right across the sky, but it was then actually shorter than it was a few weeks later when its apparent angular length was less than 90°. Even for a great comet the tail does not usually begin to develop until the comet is rather less than about two astronomical units from the sun, for most comets less than one unit, and as the distance further diminishes so the tail development correspondingly increases. As a general rule the smaller the perihelion distance the stronger the tail, other things being equal. Small comets may approach the sun much more closely before tail formation commences. For most comets, the activity within the coma and nucleus, and of tail formation, appear usually to be greater after perihelion than at the corresponding position before it, though there are exceptions to this.

The tail has the interesting property that it appears to stream behind as the comet approaches the sun, but precedes as the comet travels away again beyond perihelion. The general rule is that at different part of the orbit the tail is always streaming away in practically the opposite direction to that of the sun as seen from the comet, so that while the tail lies in the orbital plane the angle between the general direction of the tail and the direction of orbital motion of the comet varies considerably. The behaviour of the tail may be understood by the simple analogy of the smoke of a steamer blown by the wind and moving away from the vessel in a direction quite independent of the course, but where cometary tails are concerned the part of the wind is played by the repulsive pressure of the sun's radiation so that each particle of the tail-forming material probably describes a hyperbolic orbit about the sun as outer focus. The curvature of the resulting paths often gives to the tail as a whole a general slightly curved appearance, descriptively likened to the blade of a scimitar. The tail appears continuous with the outer part of the coma and grows progressively fainter and broader with increasing distance along its length till it eventually becomes no longer observable. A number of comets have actually shown several distinct tails of differing lengths and directions, some of them however often being extremely faint. Cheseaux's comet of 1744 is said to have shown six distinct tails, and the great comet of 1825 showed five, but the record is claimed for Borelly's comet (1903 IV) for which nine tails were discovered photographically at Greenwich. Also what look like short stubby tails (sometimes termed 'beards') are occasionally found jutting out small distances from the comet in quite unexpected directions, in some instances even directly towards the sun. According to Barnard, Comet 1919b discarded its tail altogether at one stage only to produce a new one. The same phenomenon was also observed for Comets Borelly 1903, Morehouse 1908, and Halley 1910, while for Brooks' comet of 1893 the tail appeared at one time to be swept away in part as though it had there been encountered by a resisting medium.

The curvature of the tail is most conspicuous for those cases when the position of the Earth happens to lie well out of the orbital plane of the comet, and conversely when the Earth lies in or near this plane the tail is seen with little or no apparent curvature. The outer edges of tails often seem more luminous than their central regions, possibly because the tail may be more or less hollow, like a horn. Such an effect might at times give the appearance of a double tail, but it would hardly account easily for multiple tails.

From a physical standpoint comets present themselves as particularly complex objects, and, except by means of their orbital elements, their individual identification is well-nigh impossible. That an object is in fact a comet is settled in the first place by its general appearance together with its rapid motion relative to the background stars, which in an hour or so is sufficient to distinguish it from a faint nebula, but its identity with any previously known comet can be settled only by its orbital elements. Its appearance is useless for this purpose. Even if we ignore their great range of sizes, no two comets appear quite the same, and an individual comet may itself change in unexpected ways almost at any stage of its motion, and appear different yet again at a subsequent appearance. Unlike a planet, which has its own characteristic unchanging appearance, it is almost impossible to settle the identity or otherwise of two comets appearing at separated intervals, except through a comparison of their orbital elements, including the instants of perihelion passage to allow for the possibility of the comets belonging to a group travelling in almost the same orbit.

This complexity also has the effect of making it extremely difficult to give a comprehensive general account of the behaviour of comets, since almost any property that may be said to apply to the majority of them will certainly be subject to exceptions or reservations where others are concerned, and either be contradicted outright or need qualification in degree. This also seriously affects at the outset the problem of constructing the theory of the origin and structure of comets, for apparently conflicting exceptions lie almost too ready to hand and may have the danger at the early stages of development of the theory of seeming to be adverse to any hypothesis.

It is of course possible that this very aspect of comets is partly responsible for the lack of development of their theory despite the great labour expended in accumulating such a vast quantity of observational data as exists for comets. It will be realized that the accepted terminology for the various zones of comets and the phenomena that they *appear* to exhibit is purely descriptive. The words, so often made use of, such as 'coma', 'nucleus', 'envelopes', 'jets', etc., while in many cases corresponding to observed features have as yet no definite theoretical basis or counterpart, and must therefore be recognized rather as useful pictorial terms than as connoting any permanent property necessarily associated with every comet.

Observed development of a comet

As already mentioned, most comets when first observed appear like a patch of faintly luminous nebulosity. For example, Halley's comet when rediscovered on its return in 1909 showed no emission spectrum at all and its luminosity was probably due at this stage almost entirely to reflected sunlight. But as a comet approaches the sun it increases in brightness much more rapidly than any mere distance effect could account for, and the character of the light also changes in that molecular emission lines appear in the spectrum, or if present already, become far more prominent. An increase in intrinsic brightness varying inversely as the square of the distance from the sun would obviously result (apart from any phase effect) if comets shone entirely by reflected light, whereas in fact the observed increase usually takes place in accordance with a much higher inverse power. It is during this stage of increasing luminosity that the nucleus makes its appearance, if at all, and the coma may in some cases at first appear to increase in size, while at about the same time tail-formation commences, sometimes beginning as a single fine streamer but gradually broadening out into a horn-shaped diverging stream. As the distance of the comet from the sun continues to diminish it has happened for certain bright comets that jets have the appearance of being thrown out on the sunward side only to be swept back almost immediately into the tail, as explained above. The general activity of changing form and brightness,

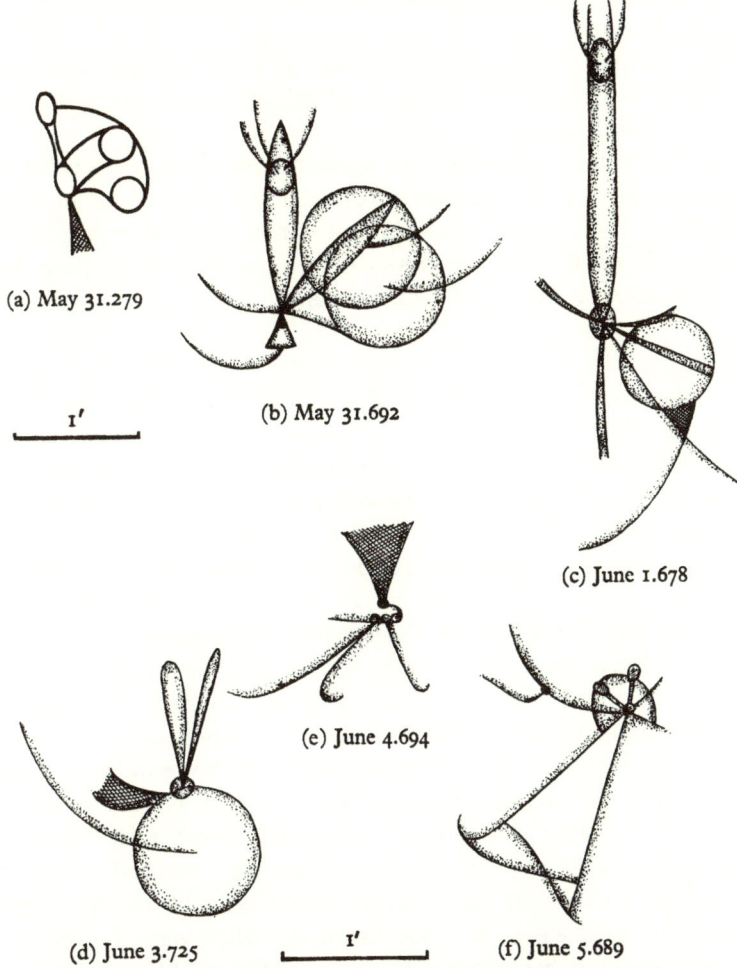

(a) May 31.279

1′

(b) May 31.692

(c) June 1.678

(e) June 4.694

(d) June 3.725   1′   (f) June 5.689

Fig. 5. From drawings by N. T. Bobrovnikoff of the head of Halley's comet (1910) made from photographs (taken at Helwan), showing the great changes in the appearance of the head that occurred within a few hours.

and of tail-production, is strongest always in the perihelion part of the orbit, but usually reaches its greatest intensity a short time after perihelion passage. As the distance from the sun increases again, activity within the comet dies down and the various effects gradually diminish in intensity until the comet becomes again a faint nebulous patch before finally disappearing. After its 1836 return Halley's comet was said to have 'vanished as if melting into adjacent space through the excessive diffusion of its light'.

This account represents only a general description of a comet's development, and departures from it may often occur in individual comets, some of which at times exhibit remarkable changes in form and brightness within a few hours at apparently quite arbitrary points of their orbits.

Sizes of comets

To estimate the geometrical dimensions of the observable portions of a comet is a simple matter when the angular diameter and distance from the Earth are known. It turns out that the volumes of a few of the largest comets are greater than that of the sun itself. Thus Holmes' comet in 1892 had an *observable* coma about one and a half million miles across—nearly twice the sun's diameter—and moreover may have extended beyond this but too faintly to be observable. This, however, was an exceptionally large and diffuse comet, and the more frequent specimens are found to have diameters ranging from about 200,000 miles to about 20,000 miles. But where the lower end of this scale is concerned, the fact that very few smaller comets are found may not at all mean that the solar system contains none with smaller diameters, for a small comet with diameter never exceeding, say, 10,000 miles could be observed only if it happened to be very favourably placed relative to the Earth and sun. Encke's comet, for example, when near perihelion, has on occasion been found to have an observable diameter of only about 3000 miles, but it has appeared a hundred times larger than this when first detected remote from the sun. Guillemin (1874) gives a list of thirteen cometary diameters showing values ranging from a mere 1200 miles for 1799 I to 1,120,000 miles for 1811 I.

PLATE III

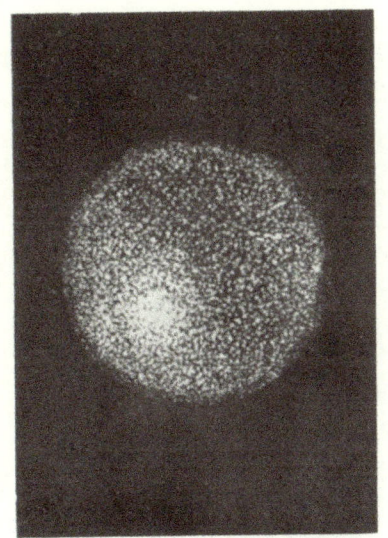

1828 November 30

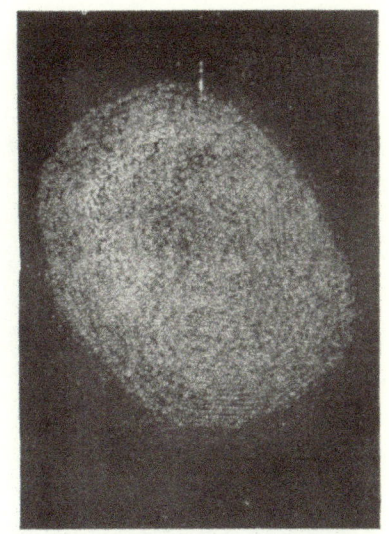

1838 August 13

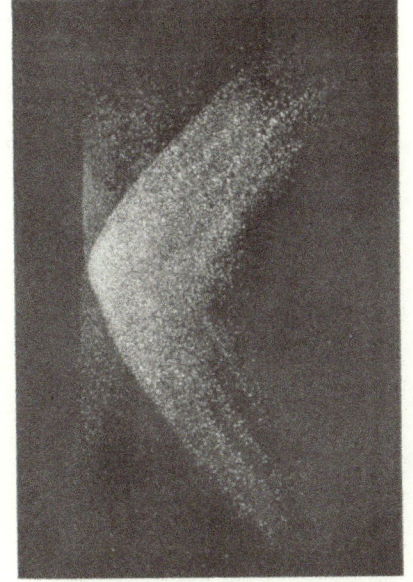

1871 November 9

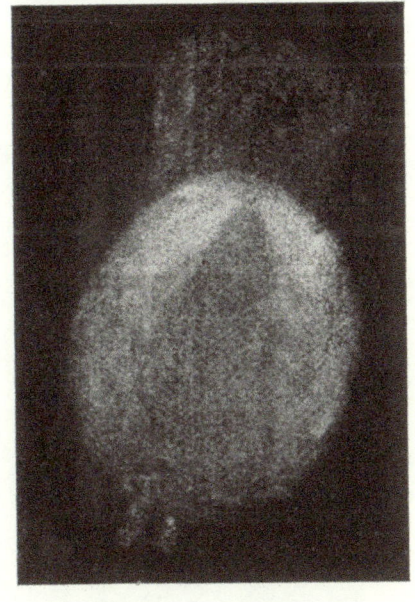

1871 December 3

ENCKE'S COMET: ILLUSTRATING ITS CHANGING APPEARANCE
AT DIFFERENT RETURNS, AND DURING THE SAME RETURN

PLATE IV

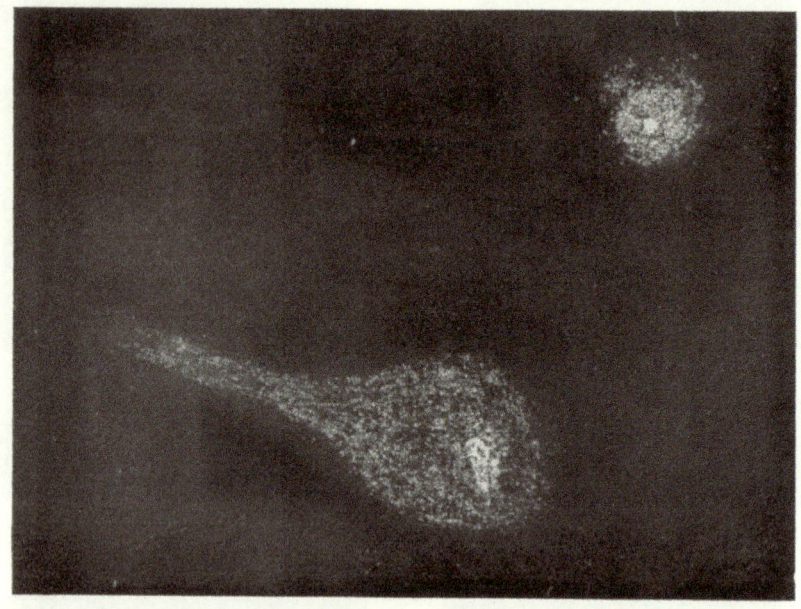

A. The components of Biela's comet shortly after disruption
had been detected in 1846

B. The components of Biela's comet at their return in 1852

PLATE V

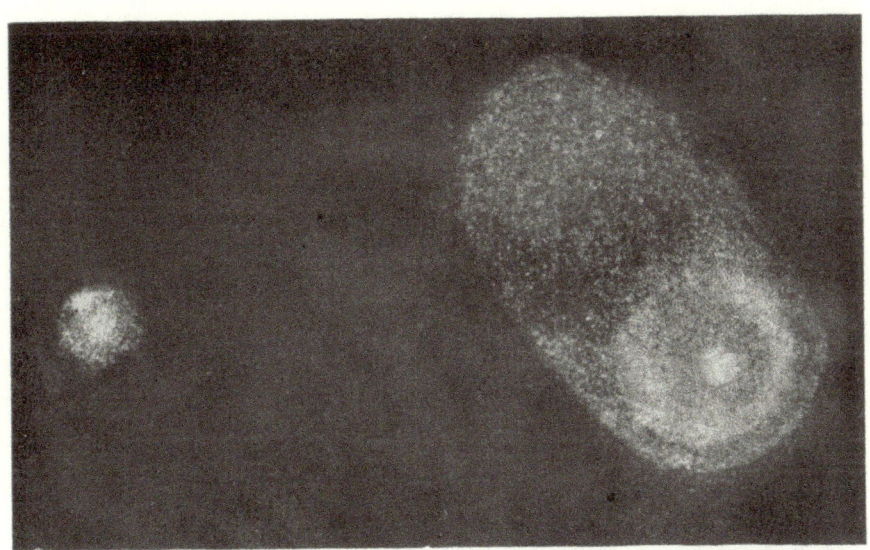

A. Sketch by Liais of the double comet Olinda, 1860 March 10

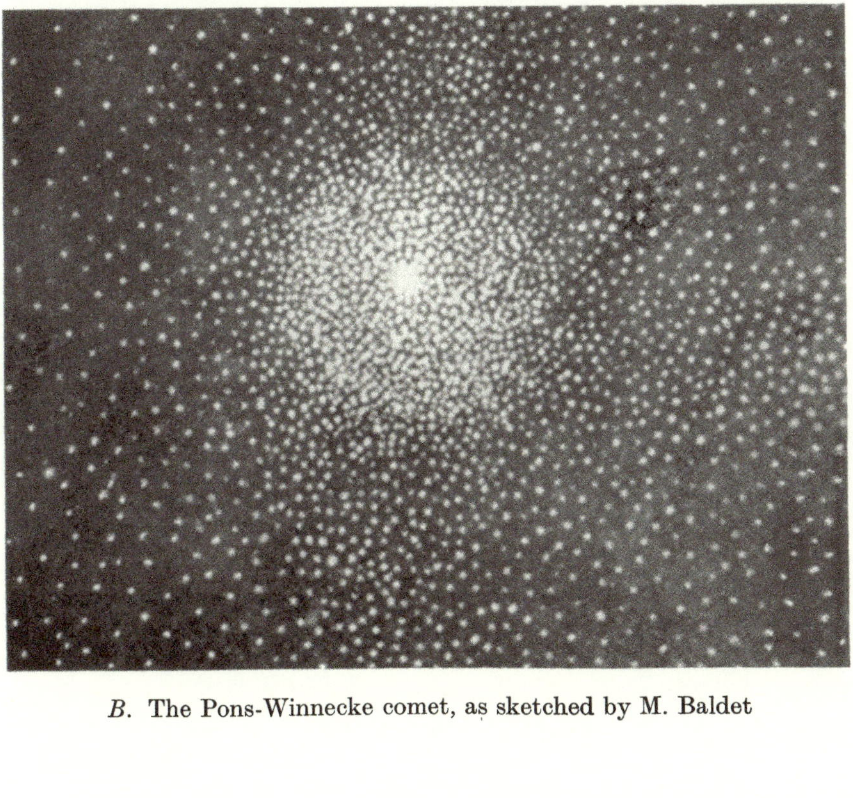

B. The Pons-Winnecke comet, as sketched by M. Baldet

PLATE VI

1910 June 4 (30-minute exposure)

1910 June 2 (5-minute exposure)

1910 May 21 (1-minute exposure)

HALLEY'S COMET 1910

This brings us to one of the most remarkable and possibly physically significant properties of comets, and that is their continual changing in size, both apparent and actual as they pursue their orbits. These changes usually show a general relation to the distance from the sun: for example, Halley's comet in September, 1909, when at about three astronomical units from the sun had an observed coma 14,000 miles across, but by the time its distance had decreased to two units the observable coma had increased to 220,000 miles. Then at perihelion, about half a unit from the sun, the observed diameter had shrunk to 120,000 miles, while later still when out at twice this distance it had increased again to as much as 320,000 miles. Finally about a year later, when its distance had increased to about four astronomical units, the observed diameter had diminished to only 30,000 miles. In considering these seemingly extraordinary changes it is important to remember once again that there is no reason to suppose a comet to resemble a planet in possessing a clearly defined and permanent outer boundary, or that the observable limit is always the same and corresponds to the actual limit of the comet. There is no evidence leading us to suppose the outer limit of the coma to be the outer limit of the comet at any stage of its development; and in addition even the observed boundary is likely to depend strongly on the conditions, favourable or otherwise, for observation of the comet, the further away the comet from the Earth and the sun, the less of its outer part may be sufficiently bright to be observable. For example, the head of the great comet of 1882 was seen to be enveloped by a very faint sheath of light extending three or four degrees ahead of it, and even this of course may not have been the absolute limit of the extent of the comet.

Such possibilities, and the likelihood of other effects, must obviously be carefully borne in mind in attempting discussion of the causes of all the curious changes of shape that have been observed for different comets. The need for such care is shown by Schroter's observations of Biela's comet in 1805, when by naked-eye measures he found its diameter to be 30,000 miles, but by telescopic observation it was only 6000. Quite apart from difficulties such as this, it has also to be remembered that the only possible direct measures are those of the projected apparent area perpendicular to the line of sight from the Earth, a direction that can bear no special relation to the comet. This has important consequences, for it means that the extent of the comet in its direction of motion when near perihelion would be difficult to measure especially if the plane of the orbit were near that of the Earth, and if the perihelion distance were less than the radius of the Earth's orbit, as is the case for over 60 per cent of comets. For, if the position of the Earth made this measurement almost transverse to the line of sight of the comet, the sun would be more or less in line with it, and would render the comet completely unobservable unless it were of exceptional brightness. Accordingly there is no reason to suppose that what may happen to be a measurable diameter can be taken as an accurate indication of the observable extent of the comet in the line of sight, if this could be observed from a comparable distance to a flank.

Hevelius, in the case of the comet of 1652, seems the first to have noticed that the absolute size increased as the comet receded from the sun. The measures on which he based his conclusion were doubted for a time, but several astronomers soon afterwards, including Newton himself, maintained that the dimensions of comets increased in proportion with their distance from the sun, and this may be regarded as a fairly generally established property of comets.

The partition of Biela's comet

Special reference may be made at this point to the remarkable developments shown by Biela's comet during its return of 1846—a year made memorable astronomically by the discovery of Neptune. This is a short-period comet (6·6 years) which at previous returns had not struck comet observers as in any way exceptional. But at the 1846 return it was seen first as distinctly elongated into pear-shaped form and then some ten days later actually divided into two separate comets which continued to travel in practically the same orbit, one preceding the other by about 175,000 miles during a period of observations extending over two months. Each of them subsequently underwent marked changes of brightness that made first one part and then the other the brighter of the two, all within a matter of a few days. At the next return in 1852 their separation had increased about eightfold, corresponding to a difference of orbital period of fifteen days, both components being readily detectable, but remarkably enough neither comet has been detected since, though the 1866 return would have provided a particularly favourable opportunity, and there can be no possibility of their having ceased to move in the same orbit. It must be that in some way they rapidly decreased in brightness after the disruption and became too faint for observation. More than one comet, although not undergoing actual disruption into separate pieces, has been found to become very diffuse, and finally disappear while still approaching the sun, notable examples being Comet Ensor in 1906, which should otherwise have become a conspicuous object, and Comet Westphal in 1913. Several instances are on record of faint cometary objects temporarily observed moving in orbits very similar to known comets but definitely displaced in time, and it seems probable that they were detached portions of the corresponding comets. For example, the very faint object 1932a, which was not seen again, was at first thought to be Comet 1932 II; while a thirteenth magnitude object was seen close to Newman's comet 1932 VII; and long before either of these recent instances there were observations of a faint object similarly associated with Biela's comet. There are also numerous instances, besides Biela's, of comets that should readily have been observed failing to be rediscovered probably as a result of disintegration, notably Comets Brorsen and Tempel, and more recently Comet Perrine in 1929.

If we may be permitted a slight digression, there was also a somewhat ludicrous side to the singular astronomical event provided by Biela's comet. The discovery of its division into two seems first to have been announced by Bradley and Herrick

43

at Yale, and by Maury at Washington, but it was also observed at Cambridge by one Challis (of questionable Neptune fame), who seems to have attributed the double appearance of the comet to a personal illusion when he first noticed it. When he again saw two comets a few days later he still thought his eyes deceived him, and not until ten days after his original observations did he feel sufficiently convinced to publish his findings. But by then it had been established elsewhere, and Challis had missed the chance fortune had allowed him of being early in the field with this discovery. Apparently Challis attributed this slowness in confirming his suspicions to the pressing claims on his attention of the search for the theoretical planet Neptune, while contrariwise later on, when hard put to it to account for his part in the latter business, he attributed his neglect of the search to the importance of his comet work. So in 1846 Challis missed both these splendid opportunities to have unique discoveries associated with his name in a way that he would have preferred, instead of perpetuating himself as an example and warning of the dangers of an ultra-cautious attitude of mind.

Until this development of Biela's comet was actually observed, the statement attributed to the Greek historian Ephorus that the comet of 371 B.C. underwent subdivision had not only remained entirely discredited but was utilized to bring its author into serious disrepute in the intervening years, showing how unsafe it is to attempt to maintain anything with certainty about phenomena not understood scientifically. A number of other early records exist suggesting disruptions of comets having been actually observed, though not sufficiently clearly described to provide really definite evidence. According to Barnard several other comets have at different times given the appearance of partial detachment of some portion, notable instances being 1860 III, the great Comet 1882 III, Brooks' comet 1889 V, Swift's comet 1899 I, Kopff's comet 1906 IV, Halley's comet of 1910, Mellish's comet 1915a and Taylor's comet 1915e. The last of these divided into two nearly equal comets in 1916 but neither was found at the return predicted for 1928. According to Young's observations at the time, the nucleus of 1882 III at first appeared circular but near peri-helion the comet elongated to a brilliantly luminous streak 50,000 miles long upon which were visible at least six star-like knots or condensations, the largest having a diameter of about 5000 miles. This 'string-of-pearls', as it was graphically des-cribed at the time, continued to lengthen, and the comet finally survived perihelion passage, where its orbit turned through about 300° in a few hours, at the price of dividing into four separate smaller comets, which all departed from the sun one after another along practically the same orbit.

In the case of Brooks' comet (1889 V), discovered in July of that year, four fragments were seen by Barnard to be separated from it early in August. Two of these were very faint and soon disappeared, but the other two had the appearance of miniature comets each complete with nucleus and tail. A month later one of these had ceased to be observable, but the other had become brighter than the main comet. At the next return (1896) only the main comet was observable.

Size of the nucleus

The nucleus of a comet usually appears to be a brighter point of light within the visible coma, and though measurements of the sizes of the nucleus have been made for different comets it is highly doubtful if such measurements can be regarded as referring to something definite within the comets. All that can probably be said is that the brightest central regions appear to have the stated extents, without any implication as to what causes the regions. In many cases the sizes are certainly near the limit of observation. A number of figures can be quoted in illustration. For example, Halley's comet in 1910 when nearing perihelion showed a nucleus 500 miles across, while the great comet of 1882 had a nucleus over 1500 miles in diameter. But these are unusually large values and Biela's comet, before disruption, is said to have shown a central condensation 112 miles in diameter with a nucleus 70 miles in diameter, according to Olivier. Then again the Pons-Winnecke comet in June 1927, when less than four million miles from the Earth, had a nucleus that seemed to Baldet at Meudon, from its brightness, to be less than 1 mile in diameter, and to Slipher in Arizona about 2 miles in diameter. Guillemin's list of diameters of cometary nuclei ranges from 28 miles to 28,000 miles, the majority being given as of order a few thousand miles. More recently Richter (1948) has listed values ranging from 40 km. for 1939c to 1000 km. for 1927i. It seems probable that the nucleus may be no more than the visible effect of a somewhat greater concentration of particles coupled perhaps with a greater depth through the comet in the line of sight at the region where the nucleus appears. Either of these two causes could obviously tend to bring about an apparent increase of brightness at a part of the coma.

Masses of comets

Despite their enormous overall dimensions and volumes, far exceeding those of any of the planets and in some cases of the sun itself, the masses of comets are quite insignificant by other astronomical standards within the solar system, and it is fairly certain that even the very largest comets cannot have masses comparable with, say, that of a moderate sized asteroid. Favourable circumstances for determining the mass of any astronomical object depend on its undergoing some kind of gravitational encounter with another body of comparable or not overwhelming mass, and evidence for the extreme small-ness of cometary masses is indirectly provided by the entire absence of perturbing effects by the comet, though the comet itself may have been strongly disturbed from its orbit. For instance, in 1770 Lexell's comet passed so near the Earth that some slight change in the planet's orbital period, that is in the length of the year, might well be expected to have been pro-duced, but the fact that not the smallest detectable change occurred showed beyond doubt that the comet's mass could not have been as much as one ten-thousandth that of the Earth. Then again, as mentioned earlier, Brooks' comet in 1886 passed through the satellite system of Jupiter and within a few radii of the surface of the planet, and thereby had its own orbit so substantially changed that the period altered from about twenty-nine years to about seven years, but not the slightest detectable disturbances of either the planet or any of its satellites were caused, and this again implies a mass far less than one ten-thousandth of the Earth. Such gravitational evidence unfortunately has provided as yet only these distinct upper limits, and it may well be that actual cometary masses are smaller by many powers of ten.

It is only by less direct theoretical arguments that closer estimates can be made and in their nature these can be only approximate, but as we shall see later, they suggest values more like 10^{-10} times that of the Earth, or say 10^{18} g., as perhaps an average cometary mass, but even such a value as this may well be a considerable overestimate for small faint comets. Recent work by Markov and others based on photometric considerations has lead to estimates ranging from 6×10^{14} g. for small faint comets to 6×10^{18} g. for large comets.

If such values are correct it becomes plain at once why no observable gravitational perturbations are produced by comets. A mass of 10^{18} g. is utterly negligible compared with the mass of any of the major planets—6×10^{27} g. for the Earth, and 2×10^{30} g. for Jupiter. But judged by ordinary terrestrial standards it is about a million million tons, which would be enough material to form a solid sphere of rock some five miles or so in diameter. If we imagined this mass spread out, in the form of smallish stones or rocks say, through a volume equal to that of the sun, it is clear that there would be plenty of empty space between adjacent stones, and if such a picture for a comet is correct it begins to give some idea of why comets appear completely transparent.

Laplace considered that the equality of rotation period and orbital period of the moon, and the exact relation $n - 3n' + 2n'' = 0$ between the mean motions of the first three satellites of Jupiter, were in themselves indirect evidence of the smallness of cometary masses, in that direct collisions between these satellites and comets must have occurred during the long history of the solar system and, unless the masses of comets were insignificant, must have disturbed these relations. The argument itself does not seem entirely conclusive, since it is made not only in the absence of a knowledge of the structure of comets but also of the explanation of these regularities, whose cause may be of such a nature as to be capable of restoring or reproducing the phenomena if only slightly disturbed.

Densities of comets

If it is assumed that the observed diameters give a general indication of all the linear dimensions of comets, it is plain at once from the foregoing estimates of masses that the average space density of matter through the whole volume of a comet must be in any event exceedingly low. And if in fact the comet extends at still lower density beyond the observed limit, as is certainly possible, the actual average density allowing for this would be still further reduced. On the other hand there is no reason to suppose that the material of the comet is spread uniformly through its extent at this extremely low density, for there can be little doubt that most of the volume of the comet is entirely unoccupied, at any rate during the inactive period away from perihelion, and that in such parts as are occupied the material is probably at normal densities consistent with the solid state of material, that is at a few grams per cubic centimetre.

But the value of the average space density is nevertheless of importance as giving some idea of the degree of separation of the material within the comet, and as indicating the possible external gravitational influence of the comet. If for the sake of numerical example we adopt, say, 100,000 km. as an average diameter and 10^{18} g. as an average mass, the density comes out to be less than 10^{-12} g. cm.$^{-3}$. This means that if such a comet were to consist of small stones each 1 g. in mass, they would be separated on the average by about 10^4 cm., or a tenth of a kilometre, rather more than a hundred yards. It would be possible to see through such a cloud of particles far more readily than through even the mildest imaginable hail. If the stones were only of unit density, the depth of such an aggregation

PLATE VII

A. Brooks' Comet 1911 November 2 (15-minute exposure)

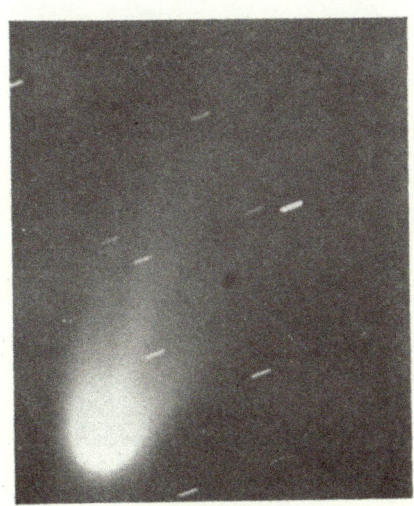

B. Delavan's Comet
1914 September 26

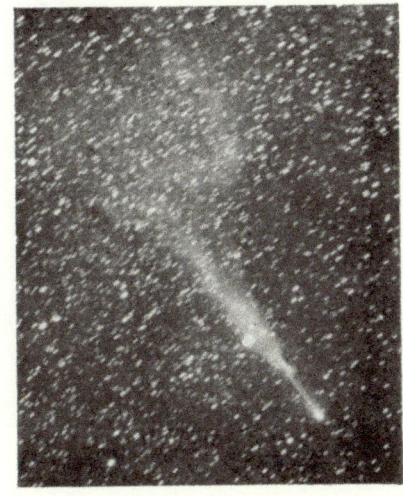

C. Morehouse's Comet
1908 October 1

PLATE VIII

A. The tail of Halley's Comet 1910 May 4 (a fortnight
after perihelion passage)

B. The head of Halley's Comet 1910 May 10 (7-minute exposure)

in the line of sight would have to reach 10^{12} cm., that is, 10,000,000 km., before the comet ceased to be transparent. If its actual depth were only 100,000 km., as may reasonably be assumed, less than 1 per cent of the apparent area of the comet would be occupied by material, and such a distribution would appear quite transparent. If some of the constituent particles were larger, or denser, the separations would be correspondingly greater and the system still more transparent.

It can be shown that a uniform, liquid or gaseous, body within the solar system will be stable against tidal disruption by the sun if the mean density of the body exceeds by a moderate factor the mean density that the sun would have if expanded so that its radius extended out to the position the body occupies. This working rule follows from the well-known approximate formula

$$D \geqslant 2(M_\odot/M)^{\frac{1}{3}} r,$$

for the distance D within which a body of mass M and radius r can with safety approach the sun. For if ρ_D is the density the sun would have if it were of radius D, then $M_\odot = \frac{4}{3}\pi\rho_D D^3$, and if ρ is the density of the other body then $M = \frac{4}{3}\pi\rho r^3$, and the above stability condition may be shown to lead to the approximate condition

$$\rho \geqslant 8\rho_D.$$

Since the actual mean density of the sun is about $1 \cdot 4$ g. cm.$^{-3}$, this leads, if the present estimate of 10^{-12} g. cm.$^{-3}$ for cometary densities is adopted for ρ, to a distance of the order of a hundred astronomical units as the order of magnitude of the limit within which such a body could approach the sun without being rendered tidally unstable. The result obviously suggests that comets may be specially prone to disruption by the sun, but it must be remembered that the argument on which the foregoing is based assumes that the notion of pressure applies to the interior of comets, and this will not be valid for a system consisting of discrete, widely separated particles. Nevertheless, since comets are observed only when well within such a limit, the result raises the possibility even at this early stage that internal gravitation for comets may be negligible compared

4

with the sun's influence, but it would be wrong simply to conclude from this that tidal disruption (whatever this may mean) of a comet must necessarily occur if its density were only 10^{-12} g. cm.$^{-3}$

All these estimates refer of course to the coma or head of the comet, but low as their densities must be, the density in the tails must be far lower still. For in the first place, the fact that the material of the tail is inevitably lost from the comet and yet many comets repeatedly produce further tails, by itself makes clear that the tail material present at any time can be only a minute fraction of the total mass of the comet. This consideration, taken in conjunction with the enormous volume even of the observed part of the tail, shows at once that the average space density is almost inconceivably small, perhaps far less than 10^{-24} g. cm.$^{-3}$ But at least some of the material of the tail may be in the form of small widely separated particles of masses of order 10^{-15} g., for these would be just the order of size of bodies that radiation pressure from the sun can strongly repel. It was shown by Schwarzschild that if the tail of Halley's comet (1910) were entirely gaseous its density may have been at most only about 3×10^{-28} g. cm.$^{-3}$ Its total mass at any one time, if wholly gaseous, need not have much exceeded 100 tons, and if entirely composed of dust possibly about 1,000,000 tons; it may well be that in actual fact it was composed of both dust and gas.

The Earth itself during recent times must on more than one occasion have passed directly through the tail of comets; it probably did so in June 1861 in the case of the great comet of that year, and again in May 1910 in the case of Halley's comet, though only near its edge. In the latter case it is certain that not the slightest atmospheric effects were present, testifying to the negligible density within the tail, though in actual fact the real densities are probably incomparably smaller than any upper limit imposed by the absence of any detectable effects. Peculiar atmospheric effects resembling aurorae were said by Hind to have occurred at the time of the 1861 passage, but these may well have been pure coincidence.

Brightness of comets

Comets exhibit a greater range of brightness than any other class of celestial objects. The brightest comets have far out-shone all other bodies except the sun and moon, from time to time highly brilliant comets having appeared that were easily visible in broad daylight. The great comet of 1882, for example, increased to such brightness shortly after perihelion that it could be clearly seen when within a few diameters of the sun (actually less than four degrees of arc away) simply by screening the light of the sun by hand. The comet of 1861, at one time with a head 30' across showed a nucleus 2" in diameter, and was brighter than Jupiter. It is estimated by Chambers that on the average four or five daylight comets appear every century. At the other end of the scale there is probably no real lower limit, since however powerful a telescope is employed faint comets are occasionally detected just above the extreme limit of observation. The periodic Comet 1933e (Wolf 1) with a magnitude then of 18·4 was probably one of the faintest comets yet observed, but in 1942 it was detected again photo-graphically by Baade when of magnitude 19·3, while 1927 IV (Stearn's comet) was followed till the spring of 1931 when it had reached the record distance for any comet while under observation of 11·5 a.u. As for changes of brightness, the great comet of 1882 itself, within a year of its perihelion passage, had become completely invisible to the largest telescopes, though its orbital position was known with all necessary pre-cision, a decrease that must have involved a change of apparent luminosity by a factor of at least a thousand million. But comets that at any stage appeared really bright have been rare, and the majority (about four-fifths) of all those discovered have remained too faint to be detected visually without aid. The number of comets visible to the naked eye averages little more than one per year.

Measures of the brightnesses of comets as they pursue their orbits make it quite clear that only a part of the variation is due directly to the changing distance (r) from the sun, as possibly the ultimate source of energy of the luminosities, and from the Earth (Δ) as the location of the observer. As already explained, instead of changing simply as the inverse square, if an empirical law $1/\Delta^2 r^n$ is assumed and the variation of brightness thrown entirely on to r^{-n}, it is found in many cases that the distance from the sun may enter roughly as the inverse third, fourth, fifth, or sometimes sixth power, even if irregular changes of brightness found in some comets are excluded. According to Baldet, the average value of n for a number of comets comes to $3\cdot32 \pm 0\cdot016$, but in individual cases at certain stages n takes a value that ranges from $-1\cdot77$ to $+11\cdot40$.

Nor is the absolute brightness of a comet the same at symmetrical positions on the two sides of perihelion. This feature was particularly noticeable for Halley's comet in 1910 when it was conspicuously brighter after perihelion passage than before, and the same phenomenon is regularly exhibited by Encke's comet. The existence of yet another complication in some comets consists of a general secular diminution of brightness, quite apart from irregular fluctuations, that seems to be associated with certain short-period comets. According to Holetschek and Vsechsviatsky, from studies of Encke's comet, a rate of decrease perhaps amounting to one magnitude per century is possibly present. This decay is doubtless connected in some way with the activity within the nucleus near perihelion, and would therefore be more likely to be detectable for short-period comets observed over many revolutions than for those of longer periods.

By eliminating the effect of different distances from the Earth and sun, comparison of the brightness of one comet with another can be made directly, and in this way it has been found that the intrinsically brightest comets of the past few centuries have been those observed in 1577, 1729, 1744, 1811 I, and 1882 III. Few other comets have been more than one-tenth as bright absolutely as any of these, and the apparent brilliance of many other comets has been due to their closeness to the Earth allied with favourable position relative to the sun. The great comet of 1729, included above, despite a perihelion distance of more than four astronomical units, was for a time faintly visible to the unaided eye, and must without doubt have then possessed an intrinsic luminosity the greatest of all comets ever seen, its absolute brightness exceeding tenfold that of any other known comet. Considering its great distance from the sun it seems highly probable that it must also have exceeded even the largest known comet in mass and size by a large factor, and if so constituted a veritable giant among comets. Comet 1925 II, whose nearly circular orbit lies entirely between those of Jupiter and Saturn and which must be a large comet to be visible at all at such distances, undergoes large irregular fluctuations in brightness sometimes amounting to as much as 5^m within a few days, and on at least one occasion as much as 9^m according to van Biesbroeck. Several other comets have shown smaller, but still very considerable, irregular changes; for example, Nagata's comet 1931 III increased from $12^{m \cdot}5$ to 8^m on 1931 October 6. There also appears to be some evidence that the brightnesses of comets of small inclinations tend to vary intrinsically to a greater degree than do those of high inclinations.

Nature of the light of comets

The light from a comet can be studied in detail by spectroscopic means in the case of those sufficiently bright to give adequate intensities when the light is dispersed. Such analysis shows that the light not only from the nucleus but also from the head and even from the tail consists in the first place of a continuous spectrum crossed by the familiar solar Fraunhofer lines originating through simple reflexion, diffraction, and scattering of the sun's light from small solid particles. The proportion of such reflected light to the whole appears usually to be greatest in the nucleus and least in the tail, and it is also considered by many observers to be greater in faint short-period comets than in bright comets, though of course in the former class the total intensity will be far less. This proportion also appears to vary according to the position of the comet relative to the sun, the comet shining chiefly, and, according to some observers, in certain instances entirely by reflected light when at great distances, but with considerable increase in direct emission as it approaches the sun; though it may be that at great distance the light as a whole is too faint for emission to be successfully disentangled. Comets with very faint nuclei, like Encke's, may at times fail to show any continuous spectrum, though this may be due simply to its being too diffuse for detection. According to Orloff, the nucleus of Halley's comet (1910) as it receded from the sun showed evidence of phase effect sufficient to suggest that it shone entirely by reflected light. Evidence of phase effect had far earlier been claimed by Delambre for the comet of 1682, and by Cassini for that of 1744. Bobrovnikoff has put forward evidence for two types of continuous spectra, one with a maximum at about $\lambda 4700$ which predominates when the comet's heliocentric distance is less than about 0·7 a.u. and which he terms the 'solar type' spectrum, and the other with maximum at about $\lambda 4000$ predominating at greater distances than this, and which he terms the 'violet type' spectrum.

The emission shows itself in the spectrum at certain wavelengths as numerous bright bands due to molecules, the particular ones actually observed for any comet depending to some extent on its distance from the sun. Identifications of several bands have been made and show that the emission is due to molecules involving mainly carbon, hydrogen, nitrogen, and oxygen, and the presence of such molecules as CH, CH^+, CH_2, CN, C_2, NH, NH_2, OH, and OH^+ has been definitely established. Comets coming particularly near the sun have also shown evidence in their spectra of sodium lines, and claims have been made for the presence also of magnesium, nickel and iron lines. But some of the very faint lines may be spurious, especially those found near the nucleus of the comet where the intensity of the reflected solar spectrum is strong by comparison. It seems probable that in the first place the molecules are released through the dissociation of more complex molecules by the action of solar ultra-violet light. Molecules such as those mentioned above when set free in this way would then be capable

54

of absorbing energy directly from the sun to re-emit it by resonance, but only molecules capable of re-emitting in the observable range of the spectrum could be disclosed. According to Swings, NH molecules can only result from photo-dissociation of chemically stable molecules liberated from solid constituents of the comet. An emission spectrum seems first to have been observed by Donati, for the Comet 1864 I, but in 1866 Huggins and Secchi found a continuous spectrum as well as bright lines in Tempel's comet. Not until 1874, for Coggia's comet, were the familiar dark lines detected crossing the continuous spectrum, enabling it to be attributed to reflected sunlight.

In the spectra of comets' tails similar emission lines are also found, and not only those due to neutral molecules but to the ionized forms as well, the most characteristic emission lines in the tails being due to CO^+, N_2^+, and CH^+. This feature is almost certainly due to the density in the tails being so exceptionally low that ionized molecules can persist for considerable intervals without capturing an electron, but meanwhile undergoing emission of energy itself originally absorbed from the solar radiation.

Owing to the faintness of comets the question whether polarization is present in their light is difficult to settle, but according to Öhman such effects are associated with the light of certain comets. According to Secchi, for the great Comet 1861 II polarization of the light from the tail and near the nucleus was strong. If this is so, it goes further to prove that comets are not self-luminous, but are dependent entirely on solar radiant energy as the ultimate source of the whole of their reflected and emitted light.

The instances of shimmering or flickering along the tail, rather similar to auroral vibrations, alleged to have been distinctly observed for some comets, apparently traversing almost instantaneously the whole length of their tails, would involve transmission of effects with speeds exceeding that of light itself, and must therefore be attributed not to any process within the comet or tail but to optical peculiarities imposed by the Earth's atmosphere.

Halley's comet at its 1910 *return*

The 1910 return of this famous comet was awaited with the greatest interest and preparation for its observation. Cowell and Crommelin had already computed its probable orbit making full allowance for planetary perturbations and had predicted the middle of April 1910 for its passage through perihelion. The actual instant was 1910 April 19·65. The first recognized observation of the comet was, however, made on 11 September 1909 by Wolf at Heidelberg, when its brightness was less than 15^m; the comet was subsequently found to have been recorded on a photographic plate taken as early as 24 August. The last observation of the comet appears to have been made on 15 June 1911, more than twenty-one months after the first. The comet was sufficiently bright to be visible to the unaided eye from 11 February to 11 June 1910, and according to Holetschek, reached a maximum total brightness of $0^{m·}7$ on 15 May, rather less than a month after perihelion.

Despite the long period over which it was available for observation the 1910 return involved numerous unfavourable features that disappointed observational astronomers. At its maximum brightness the comet was situated low down in the northern sky, while successive full moons and bad weather rendered it poorly observable for most of March and April 1910. In addition to all this the relative position of the Earth and sun while the comet was approaching perihelion made observations at this part of the orbit impossible, while at perihelion and just beyond, the line of sight from the Earth was almost tangential to the orbit. This is a common occurrence with periodic comets of small perihelion distance. Continuous observation of the tail could accordingly not be made until much later than had been hoped.

Bobrovnikoff made a study of several hundred photographs and spectrograms of the comet and produced in 1931 his discussion of this material in an extensive paper published by the Lick Observatory that ranks probably as the most thoroughgoing account of the apparition of an individual comet yet undertaken, but it is impossible here to include even an outline of this important work. It established that both the nucleus and coma showed a general contraction when approaching perihelion and expanded again afterwards, but besides this there often took place sudden changes within the nucleus, apparently of an explosive character. Secondary nuclei, so termed, were often visible, and at times the comet gave the impression of being on the verge of breaking up though without ever actually doing so. Remarkable changes within the head of the comet took place at times within merely a few hours. Photography cannot readily be utilized to study such changes because of the ever-changing details which tend to smooth out the image, but Bobrovnikoff gives a number of sketches of the appearance of the head of the comet at intervals of a few hours, and these are reproduced in Fig. 5, p. 39. They give some idea of the extraordinary changes that occur within the comet and show that the comet cannot be regarded as having any permanent form. Where the structure of the nucleus is concerned Bobrovnikoff concludes that it must consist of large numbers of separate bodies with diameters far smaller than the distances separating them. Bobrovnikoff ends his long investigation on a note of dissatisfaction with its outcome, stating that it is possible that some fundamental process within comets entirely escapes the eye and the photographic plate, and that what is observed may be but a link in the processes of interaction of cometary matter and solar energy. We shall see later that a simple dynamical process has indeed long been hitherto overlooked, that is certainly capable gravitationally of accounting for many of the observed properties of comets.

Relation between comets and meteors

A direct connexion between certain comets and meteor streams round the sun has long been established beyond any possible doubt. The idea of a close relationship seems first to have been clearly perceived in 1861 by Kirkwood, who claimed that meteors and meteoric rings were simply the debris of disintegrated comets whose material has somehow become distributed along their orbits. Some relation seems also to have been vaguely conjectured as far back as 1834 by Olmsted following the exceptionally great meteor shower of 1833. But the first theoretical evidence in support of this association was advanced by Schiaparelli, later to be celebrated for his discovery of the 'canals' on Mars, who in 1866 showed that the August meteors (the Perseids, as they are termed from the position of their radiant point), moved in orbits very closely resembling that of Comet 1862 III in both position and shape. Then in 1867 it was found that the Leonids move in an orbit practically identical in position with that of Tempel's comet (1866 I), and soon after this that the Andromedids (now also termed the Bielids) are similarly related to the orbit of Biela's comet. Numerous other identifications have since been established, such as between Halley's comet and both the May Aquarids and the Orionids, and between Encke's comet and the autumn Taurids, to mention but a few. It is highly probable that every comet has an associated meteor stream moving in much the same orbit, but obviously only those whose orbits happen to pass fairly close to the actual orbit of the Earth and have sufficiently great density are capable of making themselves known by producing meteor showers at the particular time of year concerned.

The fact that some showers recur more or less regularly each year, or for a number of successive years, at practically the same time shows that the streams of tiny particles responsible must be spread out far along the orbit, occupying a considerable fraction of its entire length. They must also be dispersed somewhat transversely to the orbit, for sometimes meteor displays may last several days, and this interval must give a measure of the time taken by the Earth to cross through the stream, though of course it may not do so centrally or orthogonally. When a meteor shower recurs regularly each year, it can be reasonably inferred that the corresponding stream must be distributed right round the cometary orbit forming a closed ring of widely spaced particles, but the displays associated with such streams are usually rather feeble. On the other hand many denser streams appear to be only partially distributed round the orbit, suggesting that they are at an earlier stage of development, but the corresponding displays may be far stronger, though successive annual showers are by no means always of comparable intensity. The Leonid shower of 12 November 1833 was so dense that at certain places a number of the order of 200,000 meteors per hour were bright enough to be observable from a single station, and for the showers of 27 November 1872 and 1885 the maximum rate was estimated at between 50,000 and 100,000 an hour, whereas a rate of only 10 visible meteors an hour emerging from a common radiant would still be considered a quite definite, though weak, shower.

The adaptation of radio technique to the problem of detecting meteors has enormously facilitated discovery, as the paths through the atmosphere of meteors smaller than those producing visible effects can be readily recorded, and the method is equally applicable to the study of daytime streams. Modern researches conducted by these methods by Lovell and his co-workers completely confirm the earlier results on the distribution of streams, but they are able to provide much ampler data on the widths and densities of streams, and the degree of dispersion along the orbits.

There is little evidence of any correlation between meteor showers and the fall of meteorites—celestial objects sufficiently massive to penetrate the atmosphere and reach the surface of the Earth before complete vaporization has taken place. A number of instances are recorded of meteoritic falls at the time of meteor showers, but these seem no greater in number than could reasonably be attributed to coincidence. Aristotle mentions that a meteorite fell in 467 B.C. when there was visible a bright comet, which is thought to have been a return of Halley's, while in more recent times, according to Young, a meteorite fell in Mexico in November 1887 at the time of the Bielids, and according to Crommelin, in 1910 a few *meteors* fell to Earth during the Aquarids associated with Halley's comet. On the other hand, although the famous 1908 Siberian meteorite fell at the time appropriate to the Pons-Winnecke meteor stream, its direction of motion, according to Crommelin, was entirely inconsistent with its being a member of the stream. If there were any direct relation between meteors and meteorites it would seem most likely to manifest itself during strong meteor showers, such as that of November 1833, though no instances of meteorites appear to have been recorded then. According to the theory developed later in this work it seems probable that comets consist of extremely small particles only, and this would imply no connexion between meteors and meteorites. Nevertheless it would be of great interest if the times of occurrence and paths of meteorites were carefully studied with a view to providing further evidence on the question.

The structure of comets

From the large amount of evidence gradually accumulated on the behaviour and properties of comets there have emerged general qualitative ideas as to their structure. In concluding this chapter we can scarcely do better than quote the considered views on this subject of H. N. Russell, than whom few can have a more detailed knowledge and grasp of the diverse phenomena exhibited through the ages by the large number of comets already known. Here then is what Russell has to say on this question (1945):

> The accepted view of the nature of comets is that they are loose swarms of separate particles, probably of very different sizes, separated by distances great in comparison with their own diameters and accompanied by more or less dust and gas. The greater part of the mass is probably concentrated near the centre of the cluster, but even here the open spaces must be exceedingly large compared with the particles.

Similar views have been reached by numerous other lifelong cometary workers such as Guillemin, Proctor, Reid, Crommelin, Plummer, Dubiago, Bobrovnikoff, Wurm, and Olivier, to mention only a few. Correct as these ideas may well prove, it must nevertheless be appreciated that they are necessarily of a somewhat conjectural nature, but the theory to be presented here adduces definite evidence in their favour. It may also be noticed that Russell's description in itself gives no explanation of the great changes of shape that appear to take place in comets, nor of the precise causes leading to tail-formation, but it will be shown in a later chapter how such things inevitably follow for simple dynamical reasons if comets are assumed to have the structure suggested by Russell. An even more interesting matter to begin with, however, is that of understanding how such curious aggregations of particles can have come to be associated with the sun and in such huge numbers, and it is to this important cosmogonical problem that we will first turn our attention in the chapter that follows, since its discussion will also provide clues to other related problems dealt with afterwards.

III

THE ORIGIN AND FORMATION OF COMETS

Introduction

The present chapter will be devoted to explaining how comets may come to be formed and also become dynamically associated with the sun. In brief, the proposed process is that of the accretion of interstellar dust through the gravitational action of the sun during passages through galactic dust clouds. It will be shown how this can result in the development of suitably compact aggregations of dust particles initially describing almost parabolic orbits round the sun. Estimates of the sizes and masses of these clusters of particles can be made mathematically from the theory of the mechanism, and also a rough calculation of the total number of such cometary aggregations likely to be brought into existence during the traverse of a single cloud. But before proceeding to the detailed discussion of the process, the analysis of which is somewhat intricate, it will be convenient to give first a short account of the evidence for interstellar dust, and the nature of its properties and distribution within the galaxy, since these things bear directly on the hypotheses of the theory.

Interstellar dust

The presence of obscuring matter in interstellar space seems first to have been recognized by Barnard, and that in fact it exists in the form of fine dust was established by Slipher from a study of the so-called reflexion nebulae often associated with dark regions. There are also numerous other observational factors that taken together have gradually confirmed the presence of such dust and further evidence is still accumulating. The frequent occurrence of dark lanes and patches, most of which are far too small and irregular in shape to be explicable as arising from the distribution of stars, provides direct evidence of material having strong obscuring power, a property highly characteristic of dust. Mass for mass, the more finely divided dust is, the greater area per unit mass can it screen, so long as the sizes of the subdivided particles do not fall much below the wave-length of the light concerned. For example, a centimetre cube of material could in the first place screen an area of 1 sq. cm. from light travelling perpendicular to one of its faces. But if it were divided into cubes 10^{-2} cm. in length and these all placed side by side, they would produce an area of 100 sq. cm. with a thickness 10^{-2} cm., and so would now screen 100 times the former area. Redistributed in this way the mass of 1 g. would now have 100 times its original screening power. Similarly if the subdivision were into cubes of side 10^{-4} cm. the effective area would be 10^4 sq. cm., and so on. For typical wave-lengths of stellar light the limit of size down to which this argument holds is about 10^{-5} cm., but beyond this the opacity begins to diminish rapidly. An everyday illustration of the immense light-stopping power of fine dust is provided by the smoke of a steamer or railway-engine, which, though sufficiently small in mass to be suspended and carried along in the air, nevertheless may be nearly completely opaque, as is shown by the dark shadows it can cast in cutting off the sun's light almost entirely.

Quite apart from the more or less directly observed patchy distributions which must be fairly near us as compared with general galactic dimensions, there is also evidence of a wide-spread general distribution of dust (probably itself the result of all the smaller scale irregular distributions throughout the galaxy), which discloses itself in the form of the well-known zone of avoidance situated in and near the galactic plane, and within the angular area of which no external galaxies or globular clusters are to be found. The distributions of these latter classes of objects over the rest of the celestial sphere are such as to render it practically certain that the 'empty' zone is only an effect produced by obscuration. There is little doubt that the same phenomenon is present in other galaxies and shows itself in the form of the dark equatorial belts observed in those spiral nebulae that happen to be seen edge on. These external systems are known to be stellar distributions comparable with our own galaxy, and the belts strongly suggest the presence of large-scale dust clouds lying in and near the equatorial planes of the systems.

There is also the well-established effect of reddening of the light of distant bright stars brought about by selective absorption, and this not only indicates the presence of dust but also enables an estimate to be made of its general density distribution, at least for those particles within the effective range of size for the process. It was early shown by H. N. Russell that all the necessary absorption could be produced, while at the same time keeping the space density well below what then seemed to be the (extremely low) limit permissible dynamically, if the particles had dimensions of the order of the wave-length of stellar light. More or less evidence for the actual sizes is provided first from the fact that no observed diffraction effects are found associated with stars embedded in nebulosity, a feature that requires the absence of particles as large or greater than about 10^{-3} cm. in diameter. A lower limit is suggested by the fact that there is found no marked difference of colour between reflexion nebulae and the bright stars that happen to illuminate them, and also by the absence of polarization effects in the light of such nebulae. Both these effects would be present if an appreciable proportion of the particles were smaller than 10^{-5} cm. in diameter. This lower limit is also confirmed by studies of the actual scattering effects found in the neighbourhood of certain O and B stars, which lead to estimated values of about $1 \cdot 5 \times 10^{-5}$ cm., the effective particles being of fairly homogeneous size with a strong maximum at this value. It does not of course follow that smaller size particles can nowhere exist in any clouds, for it may simply happen with these bright stars that the stellar light pressure has repelled them from the reflexion nebulae.

For the purpose of developing the theory of comets it is not necessary, from a philosophical or scientific standpoint, to go beyond the fact of the existence of dust particles in space, for

PLATE IX

A. 1908 September 30·873

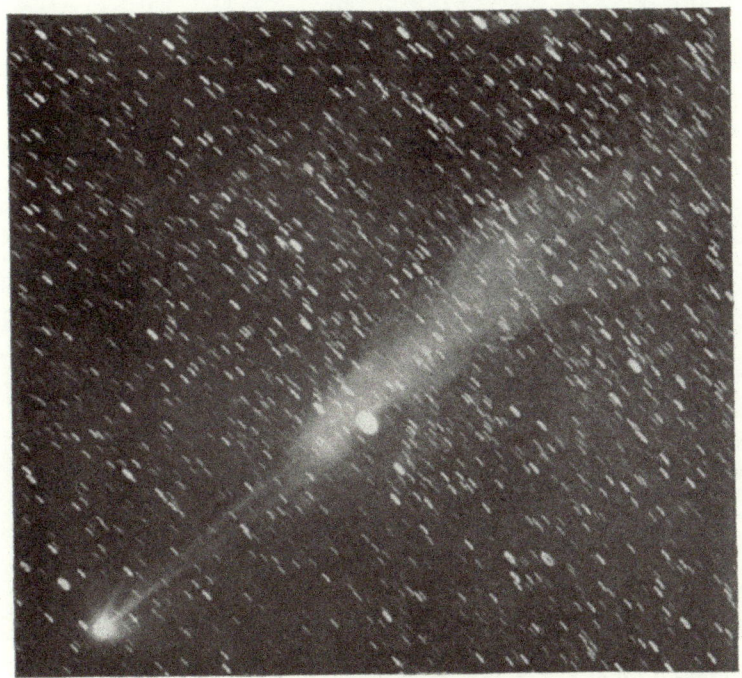

B. 1908 October 1·615

MOREHOUSE'S COMET (1908 III) SHOWING GREAT CHANGES
WITHIN THE TAIL OCCURRING IN LESS THAN 24 HOURS

PLATE X

A. Finsler's Comet 1938 August 7 (5-hour exposure)

B. Holmes' Comet (1892 III) 1892 November 10; showing almost spherical coma and no tail

C. Delavan's Comet 1914 September 20; showing a divided or multiple tail

this in itself provides us with adequate warrant to postulate them in any theoretical work, the question of their own origin being an entirely separate problem that must itself of necessity be capable of eventual theoretical solution. But even so it can do no harm to mention at this point, for the sake of completeness of the picture, that there are strong theoretical reasons for believing the origin of galactic dust to be closely connected with the explosions of super-novae, during which process large quantities of material, probably mainly in the form of heavy elements, are ejected into the surrounding space where they will later proceed to condense into small solid particles.

Though the distribution of galactic dust is very irregular, it is nevertheless concentrated in clouds, of very complex and indefinite forms, to which it is difficult to assign any definite size, though the linear dimensions may in some cases be many parsecs. The well-known dark lane in Ophiuchus, for example, has been estimated to have a projected area twenty parsecs long and about one or two parsecs wide, while a few others have far greater extents, and there are recently observed instances of small patches. The density of dust within certain clouds can be roughly estimated by consideration of scattering effects, and values of order 10^{-25} g. cm.$^{-3}$ are obtained. Though such an estimate will suffice for our purposes, it can hardly be supposed that all dust clouds are of precisely the same densities, or that a particular cloud has permanently the same density or the same density throughout, and values ten times less or greater than 10^{-25} g. cm.$^{-3}$ may well occur.

Dust clouds are usually found to be associated with the larger and more general clouds of gaseous interstellar material, itself probably largely hydrogen, which though causing negligible obscuration usually outweighs the dust by a large factor to contribute by far the major part of the mass of the cloud, and itself may have an average density of order 10^{-22} g. cm.$^{-3}$ or possibly in some cases even higher. These clouds occur more or less separately and have their own peculiar velocities in the galaxy, in much the same way as have the stars, superposed on the far more rapid general galactic rotational velocity. Because of the presence of the denser gas it can be expected that relative motions of the dust particles within the cloud are extremely small, and hence that, at any rate in any moderate-sized region of the cloud, all dust particles will have practically the same velocity, and the cloud as a whole can therefore be regarded at any time as streaming along with a uniform motion of translation.

General description of the mechanism of accretion

Let us consider next what would happen if the sun during its motion round the galaxy were to overtake or be overtaken by such a cloud of dust particles. For this purpose the motion of the sun can be regarded as rectilinear, since the time taken to pass through a cloud would be quite small compared with the time taken by the sun to complete a circuit of the galaxy. Thus the motion can be considered as though the sun were at rest and the whole dust cloud, when at great distance, moving towards it with their relative velocity, each particle within the cloud having this same velocity both in magnitude and direction. The line through the centre of the sun parallel to this direction of relative motion defines the so-called *accretion axis*, and in the postulated conditions every particle in the cloud will begin to describe under the sun's attraction a hyperbolic orbit that necessarily intersects this axis, at points beyond the sun on the opposite side to that from which the cloud is coming. Thus even if when undisturbed at great distance the dust cloud is of extremely low density, the action of the star will cause all its particles to converge towards the axis, and in so doing will bring about a strong probability of collisions between particles when they are at and near the axis. The axis appears as a kind of singularity in the motion, in that particles at one time lying on a two-dimensional area are brought, as it were, to occupy a one-dimensional length (the axis), and so we should expect there to result a space-density increase by a very large factor, though the density at the axis would in practice be prevented from being infinite because, amongst other reasons,

66

the initial distribution of material is of course not continuous but discrete.

At a collision of two particles, a proportion of their kinetic energy of relative motion will always be transformed from available dynamical energy and reappear as heat within the material, and some may be lost through work done in overcoming cohesion forces to fracture or disrupt the particles. If the speed of collision were high enough, disruption of some of the solid particles might obviously occur, and also probably the conversion of some of the material into liquid or even gaseous form. To simplify ideas, let us suppose for the moment just two particles moving from points on directly opposite sides of the accretion axis and in a plane with it, and starting originally from such positions that a collision at the axis occurs. Just before impact, it can be shown, they would have the same radial component of velocity along the axis, but their transverse components perpendicular to the axis would be equal and opposite, and it is only these transverse components that give the particles relative velocity of collision. By this means the gravitational action of the star or sun changes the relative velocity of the particles from its initial zero value when at great distance, to values that at points on the axis near the star may be large but which gradually diminish with increasing distance along the axis.

If the two particles were of the same mass, and if they were inelastic, as would perhaps be nearly the case for dust particles, then in a head-on collision the whole of the kinetic energy of relative motion would be lost, and the velocity of the combined mass immediately after the collision would be simply the former common radial component away from the sun along the axis. As the transverse component just before collision is found to be at least comparable with the radial component out to a considerable distance from the sun, it is clear that an impact of this kind, by destroying the transverse motions, can reduce the originally hyperbolic energies of the particles to elliptic values, and thus bring about their capture by the sun. Thus we have here a process that at once bids fair to resolve the great difficulty always hitherto associated with the problem of the origin of comets, namely some means whereby definitely hyperbolic speeds can become changed to elliptic ones.

In practice, collisions of all kinds will occur, ranging from head-on at one extreme to grazing collisions at the other, but obviously only those collisions tending to cause material to remain in the immediate neighbourhood of the axis will be of much importance for producing further collisions in the early stages of the process when the sun first enters the cloud.

Instead of just two particles coming from precisely opposite directions, the actual process will involve even in a small length of the axis a very large number converging from different directions distributed more or less uniformly all round the axis. For every particle the radial component when at the axis may be shown to be the same, and, it so happens, precisely equal in amount to the original speed of approach of the cloud when at great distance. Because of this, by simple relativity, the radial component of velocity can be ignored in thinking about the collisions, and the effect will be just as if particles were converging from all directions at right angles to the axis. This situation, brought about by the sun's gravitational field, will quite obviously be highly favourable for the occurrence of collisions at and near the axis.

In the initial stages when the sun first enters the cloud, the smaller the dust particles are, for a given density of cloud, the more likely will they be to collide because of their greater cross-sectional area, mass for mass, and so the more effective will they be in starting an accumulation of material round the axis. It is important to notice, however, that once such a stream at the axis had begun to form, further incoming particles would stand a much higher chance of undergoing collisions with particles already moving along the axis, and also by such means would be absorbed into the general stream.

It is clear that for particles that always remain at sufficiently great distances from the sun, the radial component velocity, which is the same before and after collision, will exceed the escape velocity from the sun at the distance of the collision, and such particles must therefore flow away from the sun in the same direction as the general motion of the cloud and eventually escape altogether. On the other hand, for particles colliding sufficiently near the sun, the residual radial velocity in the initial stages of the process will be less than the escape velocity, and such material although at first moving outwards would be drawn back along the axis by the sun's attraction and eventually move inwards. Accordingly somewhere on the axis there must develop a neutral point, as we may term it, beyond which material arriving at the axis eventually escapes, and within which it is permanently captured. Now provided the extent of the cloud is large compared with the linear dimensions involved in the mechanism of accretion (as will be shown later is the case), the process will have time to reach a steady state in which every feature is continually maintained without undergoing any fluctuations so long as the sun remains within the cloud. What in fact happens is that the stream at the axis builds up to a certain density, which varies slightly from place to place along its length, and the velocity in this stream is always towards the sun out to a certain distance, and always away from the sun beyond this distance. Particles whose hyperbolic paths intersect the axis nearer the sun than this neutral point join the inward stream and are carried towards the sun, while those entering the stream beyond the neutral point are carried away to escape in the opposite direction.

The dynamical problem of the accretion of dust

The characteristic linear dimension associated with the accretion process, and with it the distance of the neutral point from the sun, can be calculated in order of magnitude by simple dynamical considerations. For suppose that an element of volume of the cloud containing several dust particles is initially moving in a line at perpendicular distance p from the accretion axis. Assuming that collisions with other particles occur at and near a point C on the axis, the particles will not have sufficient energy to escape if their velocity afterwards, which will be along the axis, is less than the parabolic velocity at the distance SC from the sun S. The element of volume can be regarded, before it reaches the axis, as describing a hyperbola with polar equation

$$\frac{l}{r} = e \cos \theta + 1.$$

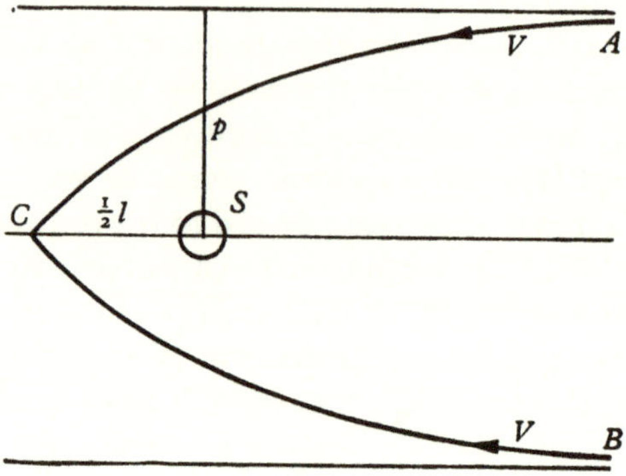

Fig. 6.

The curve coincides with its asymptote for $r = \infty$, and hence the initial direction of motion, which is along this asymptote, is opposite to that given by $e \cos \theta + 1 = 0$. Hence SC corresponds to $e \cos \theta = 1$, and therefore $SC = \frac{1}{2}l$. Also for this θ

$$e \sin \theta = \sqrt{(e^2 - 1)}.$$

Now by the law of areas $\quad r^2 \dfrac{d\theta}{dt} = h,$

the constant on the right being given by $h = \sqrt{(\mu l)} = Vp$ the angular momentum per unit mass, where $\mu = GM$ and G is the constant of gravitation and M the mass of the sun, while V is the velocity at great distance. Hence we have

$$\frac{dr}{dt} = \frac{r^2}{l} e \sin \theta \frac{d\theta}{dt} = \frac{eh}{l} \sin \theta.$$

Since p is the perpendicular from the focus S on the asymptote

$$p = \text{semi-minor axis} = (e^2 - 1)^{-\frac{1}{2}}l.$$

Hence at C we have

$$\frac{dr}{dt} = \frac{\sqrt{(e^2 - 1)}}{l} h = \frac{h}{p} = V,$$

70

which establishes the result mentioned earlier, that the radial velocity component at the axis wherever the particle crosses is always precisely equal to the initial speed of the cloud V at great distance. It follows at once from considerations of energy that the transverse component perpendicular to SC must always be precisely the parabolic velocity at C relative to the sun. Thus the collisions will decrease in violence with increasing distance from the sun.

The residual radial velocity of particles colliding at C, if we ignore all other particles at present, will be insufficient for escape if

$$V^2 < \frac{2\mu}{SC} = \frac{4\mu}{l} = \frac{4\mu^2}{V^2 p^2}.$$

Hence, on this simple picture, the particles will fail to escape after collision if

$$p < \frac{2GM}{V^2}.$$

Of course, in any actual case particles will be colliding at all parts of the axis, and some longitudinal interchange of energy, not taken account of by the foregoing simple calculations, may enter because of the different strength of the sun's attraction at different distances. But the calculation nevertheless makes clear that the characteristic dimension involved in the process is a length of order $2GM/V^2$, and a simple numerical calculation with any reasonable value for V of a few kilometres per second shows that this effective 'capture radius' is extremely large compared with the radius of the sun itself. For example, if we adopt c.g.s. units we have

$$G = 6\cdot67 \times 10^{-8}, \quad M = 1\cdot992 \times 10^{33}, \quad V = 10^5 v,$$

where v is the velocity expressed in kilometres per second, and then

$$\frac{2GM}{V^2} = \frac{3\cdot8 \times 10^5}{v^2} \text{ solar radii} \quad (1 \text{ solar radius} = 6\cdot965 \times 10^{10} \text{ cm.})$$

$$= \frac{1\cdot8 \times 10^3}{v^2} \text{ astronomical units} \quad (1 \text{ a.u.} = 1\cdot497 \times 10^{13} \text{ cm.}).$$

71

If v were even as large, say, as 10 km. sec.$^{-1}$, and smaller values are quite likely to occur, the order of magnitude of the capture radius would nevertheless be nearly 4000 solar radii, which is about 18 a.u.—nearly the radius of the orbit of Uranus. This simple analysis makes clear the enormous gravitational sweeping power of the sun arising because of the collisions at and near the axis.

The steady-state motion

The foregoing method of estimating the capture radius is entirely confirmed by the fuller investigation of the process made by Bondi and Hoyle, which in its essentials is as follows.

If we consider unit length of the accretion axis, it may be shown to begin with that in a steady state the amount of material per unit time crossing it is the same at all parts of the axis. For the amount crossing the element of length $\delta(SC)$, say, situated at C will be precisely that entering across an annular area bounded by radii p and $p + \delta p$ from the axis, that is, a mass $2\pi p \delta p V \rho$ where ρ is the density of the element when at great distance. But we have, since $p = h/V = \sqrt{(\mu l)}/V$,

$$\delta(SC) = \tfrac{1}{2}\delta l = V\sqrt{(l/\mu)}\,\delta p = \frac{V^2}{\mu} p\delta p.$$

Hence the amount crossing an element of length $\delta(SC)$ is $A\,\delta(SC)$ where

$$A = \frac{2\pi G M \rho}{V}, \tag{1}$$

and hence the same quantity of mass, namely A, crosses per unit length per unit time everywhere on the axis. Particles that may for any reason fail to undergo collisions at the axis necessary to absorb them into the general stream can be regarded as not included in the amount A, so that ϱ evidently will then refer to an 'effective density' of the cloud rather than its actual density. This might happen if there were influences within the cloud, such as small random motions, leading to particles with paths passing too far from the axis, though there is no reason to expect that the effective density will differ substantially from the actual density.

In the steady state, when particles are absorbed into the stream at the axis, they will almost at once be brought to the local velocity of the stream by means of further collisions destroying any relative velocity along the stream. In any actual case the stream must obviously have some width and extend a little way from the axis, but it will be shown later that this width is likely to be very small compared with the length of the stream, which can therefore be adequately represented by a line distribution of matter. For a general part of the stream at distance r from the sun, let us denote by m the mass per unit length in the stream and by u the stream velocity measured positive when away from the sun. If we now consider a small cylinder surrounding the stream and extending from r to $r + \delta r$ along the axis and with its end faces perpendicular to the axis, the amount of matter entering it at the end nearer the sun per unit time will be mu, while the amount leaving at the other end will be $mu + \delta(mu)$. In the steady state the difference $\delta(mu)$ must be made up by the amount entering laterally for the length δr, namely $A \delta r$, and accordingly the expression of the conservation of mass within the cylinder leads to the equation

$$\frac{d}{dr}(mu) = A. \tag{2}$$

In a similar way the conservation of momentum within the cylinder leads to the equation

$$\frac{d}{dr}(mu^2) = AV - \frac{\mu m}{r^2}, \tag{3}$$

for the only collisions tending to transfer momentum along the stream are those due to the radial component V of the material entering laterally, and compared with the stream itself (as appears later) this has such small density that its absorption into the stream can be regarded as occurring immediately. This momentum is covered by the term AV, and the second term on the right arises from the momentum contribution of the sun's field in unit time at distance r.

73

In setting up these equations it is assumed that internal gravitation of the stream is negligible in so far as the essential elements of the process are concerned. Such an assumption is of course permissible conceptually in any event, and it will be convenient to proceed in this way and consider later to what extent internal gravitation may after all be present. In fact it will be shown that it is small though rising to importance at great distance from the sun, and hence that it is possible to investigate it by means of a procedure that neglects it in the first instance. This is the standard method of applied mathematics for dealing with small additional effects whose inclusion from the outset would hopelessly complicate the problem.

Since A is a constant, (2) integrates at once to give

$$mu = A(r - r_0) \qquad (4)$$

where r_0 is a constant of integration. Its value is of special importance to the problem, for as is seen from equation (4) the stream flows towards the sun, that is u negative, for distances less than r_0, but flows away from the sun at greater distances. The value $r = r_0$ on the axis corresponds to a neutral point of the stream, as it were, and it will be shown that r_0 is always of order μ/V^2, or GM/V^2, in agreement with the result already arrived at in the preceding section.

Equations (3) and (4) can be more conveniently dealt with if expressed in terms of non-dimensional variables x, y, and z, defined as follows

$$r = \frac{\mu}{V^2} x \quad \text{and} \quad r_0 = \frac{\mu}{V^2} \alpha,$$

$$u = Vy,$$

$$m = \frac{\mu A}{V^3} z = 2\pi \varrho \mu^2 V^{-4} z,$$

so that the role of distance from the sun is taken over by x, and that of the distance of the neutral point by α, that of the velocity in the stream by y, and the line density by z. The selected units are in fact such that these pure numbers are all of order unity at the relevant region of the solution of (3) and (4). The equations for the system then become

$$y\frac{dy}{dx} + \frac{1}{x^2} = \frac{y(1-y)}{x-\alpha}, \tag{5}$$

and

$$yz = x - \alpha. \tag{6}$$

Equation (5) is non-linear and as such presents considerable difficulty in determining even the general nature of its properties. It turns out, however, that the necessary information for our purposes can be obtained by graphical study of equation (5) together with such guidance on the solution as is reasonably provided by physical considerations. Where these latter are concerned it is clear that at very great distance from the sun the degree of convergence of material towards the axis gets smaller and smaller, and the velocity in the stream must be practically equal to the general cloud velocity V, or, in terms of the new variables, that $y \to 1$ as $x \to \infty$. In any actual case although the dimensions of the cloud will be very large (at least several hundred times the capture radius $2GM/V^2$ on the average) they will not of course be infinite, and so in applying the mathematical solution there is an unavoidable element of approximation through this feature.

Very near the sun, here represented by a point centre of force, the material will be falling in freely with almost the parabolic speed, so that we must have $y \to -\infty$ as $x \to +0$. Physically therefore it is to be expected that the appropriate solution of (5) will be such that y, measuring in effect the velocity, will increase monotonically from $-\infty$ at the centre of force $x = 0$, to the value $+1$ at great distance. Thus the solution sought is one for which

$$\left.\begin{array}{l} y \to -\infty \quad \text{as} \quad x \to +0, \\[2mm] y \to 1 \quad \text{as} \quad x \to \infty, \\[2mm] \dfrac{dy}{dx} > 0 \text{ for all } x > 0. \end{array}\right\} \tag{7}$$

and

It is a curious property of equation (5) that these conditions are not in themselves sufficient to settle the solution uniquely, but they nevertheless involve the interesting and important requirement that necessarily $\alpha \geqslant 1$ for them to hold, as we shall show, and this means that the capture radius is such that r_0 exceeds GM/V^2. Moreover, as far as the problem of comets is concerned it happens that this is sufficient for our purposes.

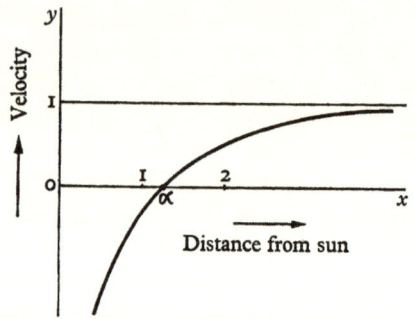

Fig. 7. Diagram showing a typical possible velocity distribution, increasing monotonically from large negative values near the sun ($x = 0$), changing sign at the neutral point ($x = \alpha$), and increasing asymptotically to the velocity of the cloud ($y = 1$) at great distance.

Before establishing the result, it is easy to show that $\alpha = 1$ partakes of the nature of a special value of the parameter α separating different types of solution. For if we rewrite (5) as

$$\frac{dy}{dx} = -\frac{x^2 y(y-1) + x - \alpha}{x^2(x-\alpha)y}, \tag{8}$$

then at a given value of x, the values of y at which $\dfrac{dy}{dx}$ vanishes are seen to be the roots of the quadratic equation

$$x^2 y^2 - x^2 y + (x - \alpha) = 0,$$

and these are $\qquad y = [x \pm \sqrt{(x^2 - 4x + 4\alpha)}]/2x.$

Thus y will be real and $\dfrac{dy}{dx}$ vanish at two real points on the ordinate of x if the quantity under the radical is positive, that is if

$$(x - 2)^2 + 4(\alpha - 1) \geqslant 0,$$

which will be true for all x if $\alpha \geqslant 1$. But if $\alpha < 1$ there will be a strip of the (x, y) plane, given by

$$2[1 - \sqrt{(1 - \alpha)}] < x < 2[1 + \sqrt{(1 - \alpha)}],$$

in which $\dfrac{dy}{dx}$ is everywhere negative.

If now in the (x, y) plane we plot the values of $\dfrac{dy}{dx}$ by means of (8), then it is readily found that the regions of positive and negative gradient $\dfrac{dy}{dx}$ in the three cases $\alpha < 1$, $\alpha = 1$, and $\alpha > 1$

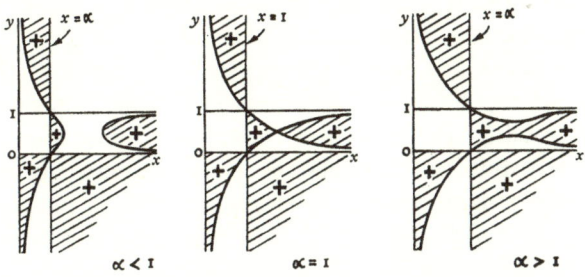

Fig. 8. Regions of positive (shaded) and negative values of $\dfrac{dy}{dx}$ as given by equation (5) for $\alpha < 1$, $\alpha = 1$, and $\alpha > 1$.

are distributed as shown in the accompanying diagrams. And if the solution is to be such that $\dfrac{dy}{dx}$ is always positive from $x = 0$ to $x \to \infty$ and at the same time $y > 0$ for $x > \alpha$, it is immediately clear from these diagrams that we must have $\alpha > 1$. For in the first case ($\alpha < 1$) there is a gap above the x-axis where $\dfrac{dy}{dx} < 0$ which could not be crossed without violating this condition. Accordingly we must have $\alpha > 1$, and this establishes that the general order of magnitude of the quantity r_0 is always GM/V^2.

If now a particular value of α greater than unity is selected, equation (6) always has $x = \alpha$ as a singular point, and every solution passing through it does so with the same gradient, namely $\dfrac{dy}{dx} = \alpha^{-2}$. Moreover, among these solutions there are an infinite number that increase monotonically *and* for which $y \to 1$ from below as $x \to \infty$. Accordingly this equation cannot suffice of itself to determine in detail what particular form of steady motion will become set up. This unexpectedly awkward property, arising from the non-linear character of the equation, gives some idea of the great complexity inherent in the general accretion problem, since in the present discussion all that we are considering is the simplified form appropriate to the accretion of dust particles. It can only be the case therefore that the actual motion that the system takes up is settled by extraneous factors, such as the particular way in which the system initially develops. This suggests that at any stage the motion is hardly likely to be thoroughly stable but rather one that will readily undergo local changes to meet any slight disturbances, but will be able to do so without vitiating the general requirement that $\alpha > 1$ together with a progressively increasing velocity from $-\infty$ near the sun to V at great distance.

Where the actual stability of a particular motion is concerned, Hoyle and Bondi have examined the possibility of random fluctuations occurring in the form of small departures from the steady state, and they conclude that even if these are present the motion will not deviate far from whatever steady state is set up provided that $\alpha < 2$. Moreover, they have shown that the precise value of α in the range $1 < \alpha < 2$, assuming always that the cloud is perfectly uniform so that disturbance from within does not occur, depends on the particular shape of the boundary of the cloud. For example, for a semi-infinite cloud with a plane boundary across which the sun enters at right angles, the appropriate value of α is about 1·25. There is, however, no reason to expect in any actual case that α would remain absolutely fixed during the passage through the cloud, and indeed the contrary is to be expected if the cloud possesses any large scale irregularity of form or density. Nevertheless the investigation shows that the value of r_0 certainly exceeds

GM/V^2 and in general will probably remain less than twice this length.

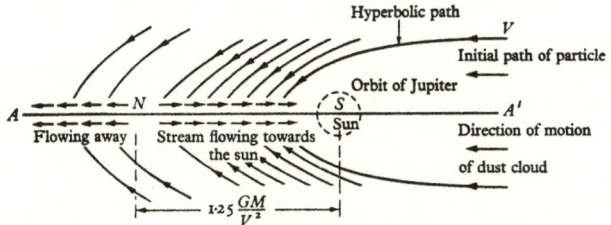

Fig. 9. Diagram illustrating the mechanism of accretion of interstellar dust by the sun.

The figure represents a section through the accretion axis AA'. The dust particles move round the sun in hyperbolic orbits and converge to the axis where a stream of material is built up. The stream flows towards the sun within a distance $1\cdot25\ GM/V^2$, and away from the sun at greater distances.

The diagram is not to scale. The distance SN would actually be of order 100 times the radius of the orbit of Jupiter (whose plane would not necessarily or usually coincide with any plane through the axis.) Also, the width of the stream would be only a small fraction of the solar radius, and is unavoidably much exaggerated in the diagram.

The density in the stream

The part of the stream corresponding to $0 < x < \alpha$ flows towards the sun and is therefore permanently captured, and it is this inward-flowing material that is of importance in the problem of comets. The line density m within the stream is proportional to z, and in absolute units is given by

$$m = z \, . \, 2\pi\rho G^2 M^2 / V^4, \qquad (9)$$

while z is related to x and y by (6), which may now be written

$$z = \frac{x - \alpha}{y}.$$

At the neutral point both $x - \alpha$ and y vanish, but the value of z can readily be found by a limiting process, thus

$$z = \frac{d(x - \alpha)}{dy} = \left(\frac{dx}{dy}\right)_{x=\alpha} = \alpha^2.$$

On the other hand, near the star $y \to -\infty$ and so $z \to +0$, while between these two points it is found that z changes monotonically. The regions that will be of special interest where comets are concerned are those within the capture distance r_0 but at as great a possible distance from the sun consistent with this, and for these parts of the stream z is found to be always of order unity, so that the actual line density in the stream may be taken to be approximately given by

$$m = 2\pi\rho G^2 M^2/V^4. \tag{10}$$

To arrive at a numerical estimate, if we take $\rho = 10^{-25}$ g. cm.$^{-3}$ and, as usual, denote by v the velocity expressed in km. sec.$^{-1}$ so that $V = 10^5 v$ cm. sec.$^{-1}$, then for the sun the formula gives a *line* density of about

$$10^8 v^{-4} \text{ g. cm.}^{-1}. \tag{11}$$

It is easily shown that such a value may not by any means imply a high *volume* density, for as we have seen, the stream must have some lateral width, and if this were as small as 10^9 cm. (less than a fiftieth of the sun's radius, and probably a considerable underestimate) the average space density would be only $3 \times 10^{-11} v^{-4}$ g. cm.$^{-3}$. It is of interest to notice, even at this early stage and with this estimated value for the width of the stream, that the volume density shows signs of being comparable in general order of magnitude with the known estimated densities of comets provided v has any reasonable value of the order of a few kilometres per second. On the other hand the value is far higher than the density in the original cloud, so that we have in the accretion process a definite dynamical mechanism that is bound to operate to produce in the first place such a stream of material at far higher density than in the cloud and flowing in a straight line towards the sun.

An extreme lower limit to the width of the stream can be obtained by assuming its material to be packed solid (though of course there is no suggestion that this would actually occur even approximately). If the specific gravity of the dust particles is taken as 2·5 g. cm.$^{-3}$, the least possible radius ϖ_0, say, is given by equating the mass per unit length to (11), thus

$$2·5\pi\varpi_0^2 = 10^8 v^{-4}$$

or $$\varpi_0 = 3·6 \times 10^3 v^{-2} \text{ cm.}$$

But as we shall see later, other effects are likely to be present causing the radius of the stream to exceed such a value by a great deal, though without making it so wide as to render inapplicable the idea of a line distribution. It has been seen that r_0 will be of order 10^{16} cm., so that even were the stream to be 10^{11} cm. wide it would yet have negligible width compared with its length and therefore be capable of being regarded as a line distribution.

Conditions for capture of particles in solid form

As each particle enters the stream once this has been built up, the velocity of collision may be as large as the vector sum of the velocity of the particle relative to the sun and the velocity of the stream at the point concerned. The former has components of amount V directly away from the sun and $V\sqrt{(2/x)}$ transversely (this component is always precisely the parabolic speed because V is the original velocity of the cloud), while the velocity of the stream is Vy. The relative velocity at entering is therefore $V\sqrt{[(1-y)^2 + 2/x]}$, and for $\alpha = 1·25$ the function under the radical has the following values:

x	1·25	1·00	0·75	0·50	0·25
$\theta = (1-y)^2 + 2/x$	2·60	3·50	5·10	8·00	15·80

If the particles are to remain largely in solid form despite the collisions, the energy converted into heat must not be such as to cause complete vaporization. If a proportion f of the relative kinetic energy per unit mass, $\frac{1}{2}\theta V^2$ (where θ is as tabulated above), is converted into heat, and if E is the energy per unit mass required to vaporize the material, the condition it should not be vaporized entirely is

$$V^2 < 2E/\theta f.$$

81

The factor f for any particular particle will depend on the way it happens to be absorbed into the stream, and since, in general, more than one collision may be involved in stopping a particle it is not possible to assign a precise value, even as an average, for f. However, a value of one-quarter would probably be an overestimate and so an unfavourably high figure for present purposes, but as it enters only to the one-half power in its effect on a calculated limit for V, this may not lead to any serious error.

Where E is concerned, according to a recent calculation by Hoyle, the heat of evaporation of iron at boiling point is about 1000 cal./g. Dust particles will not of course necessarily consist even mainly of iron, and also they will presumably have low initial temperature as compared with their boiling points, so possibly E may roughly be taken as 2000 cal./g., or about 10^{11} ergs/g., to make some allowance for both these considerations.

In that case, near the neutral point where $\theta = 2.6$, this leads to an upper limit to V of about 6 km. sec.$^{-1}$, while if complete vaporization is not to occur, say halfway along the inward stream ($\theta = 5.7$), then V must not exceed about 4 km. sec.$^{-1}$. If α were as large as 2, these values would be increased by nearly a kilometre per second for each place. In view of the uncertainties attaching to both E and f, these results go no further than to give a rough indication of possible upper limits to the speed, but nevertheless they are of the order of magnitude of velocities reasonably likely to occur in an encounter between a star like the sun and dust clouds. On the other hand, even if a certain amount of vaporization occurred, there is always the possibility of the material condensing again to reform into fresh dust particles, but this is a point that will present itself again later on.

The object of the foregoing rough calculation is to show that vaporization effects do not place any serious restriction on the process. At the points where uncertainty in the quantities concerned has been involved we have endeavoured to choose adverse values tending to depress the eventual upper limit of V, so the value of 6 km. sec.$^{-1}$ arrived at is likely to be an under-estimate. Had the result come out to, say, less than a tenth this value, it might have suggested, in the absence of a fuller investigation than at present seems possible, that the process needed exceptionally small though not impossible conditions of relative velocity. But as it is the result scarcely imposes any serious restriction on the cloud velocity.

The occurrence of collisions in the accretion process

So far it has simply been supposed on general grounds that collisions at the accretion axis will necessarily occur, whereas an original density of 10^{-25} g. cm.$^{-3}$ for the cloud is so low that it implies very large initial distances between the individual dust particles even if these are themselves very small. For if r cm. is the radius of an average particle, supposed spherical, and a cm. the average separation, then these will be related approximately by

$$a^3 \cdot 10^{-25} = \tfrac{4}{3}\pi\sigma r^3,$$

where σ is the specific gravity of the material in the dust particles. For $\sigma = 2 \cdot 5$ g. cm.$^{-3}$ this leads to

$$a/r = 5 \times 10^8,$$

so that even if r were of order 10^{-4} cm. the particles would on the average be about half a kilometre apart. The probability of collisions in a length of path comparable with the linear dimensions of a comet, for example, in a density distribution such as this would be negligibly small; for a number of particles of the order of r^{-2} would be needed to screen 1 sq. cm., and these would have a total mass $\tfrac{4}{3}\pi\sigma r$, or say, roughly $10r$ g. At density 10^{-25} g. cm.$^{-3}$ a column 1 sq. cm. in area would need to have a length $10^{26}r$ cm. to contain this mass, and for $r = 10^{-4}$ cm. this is some thousands of parsecs.

In the early stages when the sun first enters such a cloud the factors involved in bringing about collisions at and near the axis can be investigated by considering how an element of volume moving with the material, to borrow a phrase from hydrodynamics, changes during the motion. For example, let us consider an element of volume of the cloud, before it has reached the sun, having annular shape between radii p and $p + \delta p$ from the axis (where p is of the same order of magnitude as r_0), and of small length $2b$ parallel to the axis. By this is meant the ring-shaped volume got by rotating the area $ABCD$ (see Fig. 10) round the axis, the point D lying sufficiently ahead of B initially for them to arrive at the axis at the same instant.

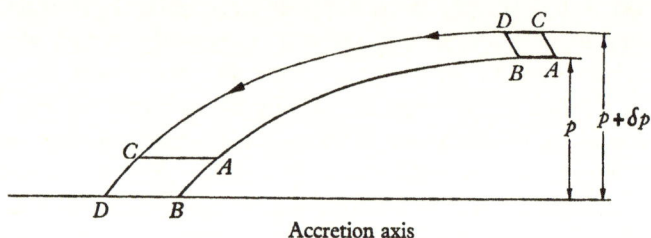

Fig. 10. Showing the changing shape of a converging element of volume.

The area of this plane figure is $2b\,\delta p$ and hence the total volume of the annulus is $4\pi bp\,\delta p$. If we now imagine each point of this element of volume to describe the hyperbolic path that a particle initially at the point and moving with the velocity V would describe, then under such a motion of all its points the annular region will converge towards the axis, steadily contracting its radius. But its other dimensions will change only by factors of order unity, since the velocity remains always of order V. Hence in this way an enormous decrease of volume will occur and particles lying within the original volume will be brought far closer together.

To calculate the extent of this compression, if we consider a point of the original volume that crosses the axis at distance R from the sun, its transverse speed there is $\sqrt{(2GM/R)}$, and hence by conservation of angular momentum its initial perpendicular distance p from the axis is given by

$$p = \sqrt{(2GMR)}/V.$$

Thus by differentiation, a length δR on the axis corresponds to an initial difference δp in distance from the axis given by

$$\delta p = \sqrt{(GM/2R)} \cdot \delta R / V.$$

In calculating the shape and volume that the element assumes near the axis the radial velocity component V, which is the same for every point, can be ignored, as already explained, and each point of the volume regarded as approaching the axis perpendicularly. Near the neutral point $(R = 1 \cdot 25 GM/V^2)$, to take some definite distance, the transverse velocity will be $\sqrt{(1 \cdot 6)} V$, and therefore the length of the element AB, originally $2b$, will have increased slightly to $2\sqrt{(1 \cdot 6)} b$. The volume element when at its smallest will have the form of a circular cylinder of radius $\sqrt{(1 \cdot 6)} b$ and length δR surrounding the axis. The least volume is therefore effectively $1 \cdot 6 \pi b^2 \delta R$.

If now the original volume element contained the centre of just one particle, then collisions will certainly occur at the axis if this reduced volume is less than the effective volume of the particle. If the particle is a sphere of radius r and its centre is somewhere within the minimum volume element at the axis, the conditions it will not overlap any other particle near the axis are

$$\delta R \geqslant 4r \quad \text{and} \quad \sqrt{(1 \cdot 6)} b \geqslant 2r.$$

If we use the limiting values given by the equals signs, the original volume element expressed in terms of the size of the particle gives

$$4\pi b p \delta p = 8\sqrt{(10)} \pi r^2 GM/V^2.$$

Collisions will therefore occur almost with certainty if this volume originally contained more than one particle of the adopted size. But if ρ is the density of the cloud and σ the specific gravity of the material of the particles, the average volume per particle is $4\pi\sigma r^3/3\rho$, and for collisions to occur more or less with certainty for every particle from the very outset of the accretion process this must be less than the former volume. With $\sigma = 2 \cdot 5$ g. cm.$^{-3}$ and writing $V = 10^5 v$ cm. sec.$^{-1}$ so that v is the speed in kilometres per second, the condition reduces almost exactly to

$$\rho > 10^{-17} v^2 r \text{ g. cm.}^{-3}$$

If, for example, r were as small as 10^{-7} cm. and $v = 1$ km. sec.$^{-1}$ then collisions will occur with a probability approaching unity for every particle from the outset if $\rho = 10^{-24}$ g. cm.$^{-3}$. For the value of ρ adopted earlier, namely 10^{-25} g. cm.$^{-3}$, this probability would be reduced to $0 \cdot 1$, and for particles of general radius r the probability would be of order $1/10^8 r v^2$. As would be expected, for a given density, collisions are more likely to occur the smaller r is, other things remaining unchanged.

When the sun first enters the cloud the smallest particles will be the most effective in starting collisions near the axis, and although the probability for any particular particle may be small it is nevertheless a finite quantity and there are so many particles that some collisions must occur practically straight away, which would tend to leave material moving along or near the axis and start the accretion process off. Once this happens collisions of further incoming particles will no longer rely on the present process, but will take place with the material already at the axis and will proceed with increasing probability soon amounting to certainty. To begin the process it is only necessary for there to be a small but sufficient proportion of the mass of the cloud in the form of particles of small linear dimensions. It may well be that the presence within the cloud of gas, especially if at much higher density than the dust, as is likely to hold in many clouds, would greatly assist the initial stages of the process, but on the other hand there seems no need to regard this as necessary to the theory.

So far we have regarded the convergence of particles towards the axis as taking place under an inverse square law of attraction to a fixed central point, whereas in fact the presence of the planets must necessarily render this only an approximation, though it must nevertheless be a closely accurate one. It will be shown later that the resulting perturbations from strictly hyperbolic paths are extremely small compared with the length of the inward moving accretion stream. But small as the deviations are, they will prevent the paths of at least some of the particles from precisely intersecting the geometrical accretion axis, and in the initial stages, when the sun first enters the cloud, this might at first sight seem to operate to reduce the probability of collisions taking place, but in actual fact it makes negligible difference to the initial probability. This can easily be seen as follows.

Suppose that we denote by s the general magnitude of planetary perturbations for particles crossing (near) the axis at a distance of order GM/V^2 from the sun. These perturbations will have components both along and perpendicular to the axis, but if we consider a small cylinder of length s and radius s situated symmetrically round the axis at the part concerned, the total number of particles entering it will be effectively the same as in the idealized undisturbed motion, since its size is comparable with that of the perturbations. This means that the number will be the same as would meet a length s of the axis in undisturbed accretion. Thus the amount of material entering the cylinder per unit time will be $2\pi\rho GMs/V$. Also, since the time taken by a particle to cross the cylinder will be of order $2s/V$, the mass within the cylinder in the initial stages of the process, when most of the particles will cross freely through the cylinder, would be $4\pi\rho GMs^2/V^2$. Since the volume of the cylinder is πs^3, the average volume density within the cylinder would be $4\rho GM/V^2 s$, and the particle density would therefore be $4nGM/V^2 s$, where n is the particle density at great distance from the sun in the undisturbed cloud. The length of path described within the cylinder by any particle crossing it will be of order $2s$, and hence if A is the collision cross-section of a typical particle, the probability of its undergoing a collision within the cylinder (calculated in the usual way on the assumption that the probability is considerably less than unity) is the product of the particle density with the effective volume, $2sA$, traversed by a particle in crossing the cylinder. This leads at once to

$$(4nGM/V^2 s).2sA = 8nAGM/V^2,$$

and *is therefore independent of s*. This accordingly establishes that small perturbations, whether produced by the planets or any other factor such as small velocity differences within the original cloud, will not affect the initial probability of collisions, and that the building up of the stream will proceed regardless of such influences.

That this result is equivalent to the earlier result derived on the basis of more or less accurate intersection of the orbits with the accretion axis can easily be seen. For, taking the particles to be spheres as before, we have $n = 3\rho/4\pi\sigma r^3$ and $A = 4\pi r^2$, and the present expression for the collision probability in the initial stages becomes $24\rho GM/\sigma r V^2$, and inserting numerical values this reduces to $10^{17}\rho/rv^2$ approximately, in complete agreement with the former result. Accordingly we may conclude that planetary or other sources of perturbations tending to deviate the particles from paths accurately intersecting the accretion line have no effect on the probability of collisions. This feature of accretion is obviously of special importance, for it means that the process is highly efficient and independent of any specially advantageous circumstances for its commencement.

When the stream has been built up and the steady state reached, it is to be expected in any actual motion that it will have a small but finite width perpendicular to the axis, for any influences tending to disturb the ideal motion, and such are bound to be present, must lead to some lateral distribution of the stream. For instance, small internal motions of the dust particles within the cloud would mean for some particles that the planes of their hyperbolic motions would not pass accurately through the accretion axis. Instead of intersecting in a single line, the various planes of motion would intersect in a whole bundle of lines passing through the centre of force and distributed round the theoretical accretion axis, but thinning out in frequency with increasing distance from the axis. The width of the stream would be related to this distribution, but obviously it would not necessarily be a perfectly definite quantity indicating a sharp boundary, for the stream must at some stage gradually thin out sideways till its density merges with that of the cloud, though the change from high to low density might take place over quite a small distance. Again, slight differences of density within the cloud must cause some lateral spread of the stream, because then the total momentum of the converging material would not balance out precisely to zero. But in this connexion it is important to remember that in the steady state the amount of material per unit length of the stream existing at any time would require a time of order GM/V^3, or about $4 \times 10^3 v^{-3}$ years, to be transported in laterally, and hence that irregularities of density, even say a complete absence of material on one side, over distances taking a few years to be covered by the motion of the sun, would produce only extremely small sideways deviations of the main stream.

For reasons such as these, and also the effects of any other kinds of disturbance, the stream will necessarily possess a finite but small width at any particular distance from the sun. Particles within the cloud, having random velocities larger than a certain amount, if such exist, might fail to follow paths passing through the stream, and instead of being intercepted would then sweep on along their hyperbolic paths and escape again without loss of energy. But a finite proportion of the material passing within the capture radius will always be absorbed into the stream. In this way the particles of sufficiently small random velocity within the cloud will be sorted out, as it were, and captured. The quantity ρ will evidently refer to the density counting only these particles, but there is no reason to suppose that it will differ much from the actual density, and as estimates of the latter are known only very approximately it will not be worthwhile or indeed necessary for our purposes to attempt to make any distinction between them.

An estimate of an upper limit to the width of the stream can be made, as far as order of magnitude is concerned, by considering how diffuse it would be possible for the stream to be while at the same time allowing the newly arrived particles only a small probability of passing through it without being absorbed by collisions. If we denote by ϖ the radius of the stream consistent with this and ρ_s its volume density, then equating the resulting line density to that already found, (11), gives

$$\pi\varpi^2 \rho_s = 10^8 v^{-4} \text{ g. cm.}^{-1}. \tag{12}$$

If now r is the radius of a typical particle of both the cloud and the stream, a particle in travelling a distance 2ϖ across the stream would drill out a cylinder of volume $2\pi r^2 \varpi$. Such a volume in the stream contains mass $2\pi r^2 \varpi \rho_s$ and will contain at least one particle of density σ if this exceeds $\frac{4}{3}\pi r^3 \sigma$, which gives the condition

$$\varpi \rho_s > \tfrac{2}{3}\sigma r. \tag{13}$$

From (12) and (13) we then obtain for the radius of the stream

$$\varpi \leqslant 1\cdot 5 \times 10^8 / \pi \sigma r v^4 \text{ cm.} \tag{14}$$

and for the density

$$\rho_s \geqslant \pi \sigma^2 r^2 v^4 / 2 \cdot 25 \times 10^8 \text{ g. cm.}^{-3}. \tag{15}$$

If we adopt $\sigma = 2\cdot 5$ g. cm.$^{-3}$ these lead to the following numerical limiting values for the stated values of the average radius r.

TABLE II

r	ϖ	ρ_s
10^{-5}	$1\cdot 9 \times 10^{12} v^{-4}$	$8\cdot 7 \times 10^{-18} v^4$
10^{-4}	$1\cdot 9 \times 10^{11} v^{-4}$	$8\cdot 7 \times 10^{-16} v^4$

For values of v of a few kilometres per second these quantities are comparable with the dimensions and densities of comets. In Chapter II we saw that comets have linear sizes ranging by at least a factor of 10 above and below 10^{11} cm., while their densities, far less certain, possibly range by a factor of at least 10^2 above and below 10^{-13} g. cm.$^{-3}$.

It may be noticed also at this point that the mean free path of a particle for motion along the stream will also be of order ϖ at most, and in comparison with the total length L, say, of the stream, namely $\alpha G M / V^2$, is always extremely small. Thus for $r = 10^{-5}$ cm. the ratio ϖ / L is readily found to be about $10^{-4} v^{-2}$, while for $r = 10^{-4}$ cm. it is still less at $10^{-5} v^{-2}$. This result shows that even though the density in the stream, ρ_s, may be very small, collisions for motion of particles *along the axis* have a sufficiently high probability of occurring for the continuous equations of motion to be closely applicable, as assumed at the outset in setting up the equations of motion for the problem.

The formation of comets within the accretion stream

The application of the accretion process has shown how a longitudinal stream of dust particles at reasonably high density must come to form when the sun passes through a dust cloud. Confining our attention to the part of this stream flowing towards the sun, let us now consider the extent to which self-gravitation of the material within the stream, hitherto neglected, compares with the effect of the sun's gravitational field. This latter field diminishes as the inverse square with increasing distance from the sun, but what is of importance within the stream is not the direct field but the differential field as between one point and another, and over a given small length this decreases as the inverse cube. Obviously then the further out a point is from the sun the more likely is internal gravitation to rise to importance, so for the inward stream we may start by investigating conditions near the neutral point. The effect of the internal attraction acting alone would be to cause the stream to pull itself together lengthwise, but the tendency of the sun's field for any given segment of the stream, by exerting a greater pull at the nearer end than the further one, is to extend its length. The possibility of the separation of the stream into a large number of segments can be examined by expressing the condition for the internal gravitation to prevail over the disruptive influence of the sun. The existence in the stream of slight centres of attraction about which such aggregations would tend to collect can certainly be assumed, for it has already been seen that there might well be small irregularities of density in the cloud, while in addition such centres might arise from the general unstable nature of the accretion process itself. Supposing such centres to exist, then were it not for the sun, the stream would begin to separate into a number of finite segments each contracting under its own self-gravitation, and the presence of other segments would certainly not assist the contraction within any particular one of them. Accordingly in obtaining the condition for contraction we may consider simply the self-gravitation of a segment itself opposed

91

by the differential field of the sun. The total mass of the stream is negligible compared with that of the sun, and the disruptive effects of the parts immediately adjacent to any segment can be neglected in making an order of magnitude estimate.

On this basis, suppose the initial length of such a segment within the stream, before serious contraction occurs, to be d. Its total mass will be of order md, actually $\alpha^2 md$ near the neutral point, so that its self-gravitation will be underestimated if we take the mass as md and this will make ample allowance for any possible disruptive influence of the nearer parts of the rest of the stream. The force of attraction at points near the ends of this segment will be approximately $Gmd/(\tfrac{1}{2}d)^2$, or $4Gm/d$. If the distance of the centre of the segment from the sun (mass M) is R, a length that may be assumed to be very much greater than d, the difference between the force at the centre, namely GM/R^2, and at the ends, namely $GM/(R \pm \tfrac{1}{2}d)^2$, is practically GMd/R^3 and acts to stretch the segment, as it were. The particles of the segment lying nearer the sun are more strongly accelerated than those at the centre of the segment by this amount and so tend to be drawn on ahead, while those at the farther end are less accelerated and lag behind. But the internal field will prevent this occurring if

$$\frac{4Gm}{d} > \frac{GMd}{R^3},$$

that is, if

$$d < 2\sqrt{(mR^3/M)}. \tag{16}$$

Accordingly the mass of the segment must satisfy

$$md < 2\sqrt{(m^3 R^3/M)}. \tag{17}$$

With the value $m = 10^8 v^{-4}$ g. cm.$^{-1}$ already found for the line density and with R measured in astronomical units, and the velocity in kilometres per second, as usual, these conditions become

$$d < 2 \cdot 6 \times 10^7 R^{\frac{3}{2}} v^{-2} \text{ cm.} \tag{18}$$

and

$$md < 2 \cdot 6 \times 10^{15} R^{\frac{3}{2}} v^{-6} \text{ g.} \tag{19}$$

92

Since the radial width of the stream ϖ will usually be smaller than but of the same order as d, as Table II (p. 90) shows, the transverse self-attraction of the stream will be of at least comparable strength to the lengthwise attraction, and will therefore also operate to cause the primitive aggregations to form.

If we consider a segment starting to develop near the neutral point, so that $R = 1\cdot25\,GM/V^2 = 4\cdot4 \times 10^3 v^{-2}$ a.u., then (18) and (19) lead, in round figures, to

$$d < 8 \times 10^{12} v^{-5} \text{ cm.} \tag{20}$$

and
$$md < 8 \times 10^{20} v^{-9} \text{ g.} \tag{21}$$

For closer distances the corresponding values are less than these by a factor that depends on the three-halves power of the distance, so the foregoing values represent upper limits to the initial lengths and masses of the segments. No definite lower limits can be decided, or are likely to exist, because there is no lower limit to R, apart from zero. Since m is proportional to $2\pi\rho G^2 M^2/V^4$, it follows that for different values of the density ρ, the value of d changes like $\rho^{\frac{1}{3}}$, while the mass md changes as $\rho^{\frac{1}{3}}$.

It is the purpose of the theory to identify these segments with the initial forms of comets. We have here a process depending on gravitational effects together with collisions by which in the first place the sun will form the stream by accretion, and then through internal gravitation this stream will divide lengthwise into segments. For any reasonable value of v of a few kilometres per second, the foregoing values of m and md are seen to be in as close agreement with estimates of cometary dimensions and masses as could reasonably be expected in view of the uncertainties attaching to the observed data on the one hand and the unavoidable approximations of the theoretical calculations on the other. The table on p. 94 shows the values of the maximum cometary dimensions and masses calculated from (20) and (21) for a number of values of v. The values of d and md depend of course also on α, and as this itself may vary from 1 to 2 according to the position of the neutral point, a further factor of $2\sqrt{2}$, or say about 3, can enter from this feature.

The total mass of the stream at any instant is of order GMm/V^2 or about $1{\cdot}3 \times 10^{24} v^{-6}$ g., and this alone would suffice for about $2 \times 10^3 v^3$ comets of the maximum estimated mass. But in view of the unstable nature of the accretion process it is to be expected that comets of all sizes, from the maximum downwards, will form initially in the stream.

TABLE III

$\rho = 10^{-25}$ g. cm.$^{-3}$					
v (km. sec.$^{-1}$)	1	2	3	4	5
d (cm.)	8×10^{12}	2×10^{11}	3×10^{10}	8×10^9	3×10^9
md (g.)	8×10^{20}	2×10^{18}	4×10^{16}	3×10^{15}	4×10^{14}
$\rho = 10^{-24}$ g. cm.$^{-3}$					
v (km. sec.$^{-1}$)	1	2	3	4	5
d (cm.)	2×10^{13}	6×10^{11}	9×10^{10}	3×10^{10}	9×10^9
md (g.)	2×10^{22}	6×10^{19}	1×10^{18}	9×10^{16}	1×10^{16}

The rate of contraction of the segments

It is of importance to consider the possible rate at which the segmentation of the stream would take place, because the accretion process depends on a more or less continuous stream being present, and so the rate of contraction should not be so rapid as seriously to invalidate this assumption.

For a segment of length d the time taken to contract substantially under its own field will be of the same order but less than the time of free fall from distance $\tfrac{1}{2}d$ to a point mass md, and this by Kepler's third law is readily found to be $\tfrac{1}{8}\pi d(Gm)^{-\frac{1}{2}}$. But by means of (16) this time is itself less than $\tfrac{1}{4}\pi R^{\frac{3}{2}}(GM)^{-\frac{1}{2}}$.

On the other hand the time of fall to the sun from a distance R is, again by Kepler's third law, $\tfrac{1}{4}\pi \sqrt{(2)}\,R^{\frac{3}{2}}(GM)^{-\frac{1}{2}}$. But the actual time of fall will be longer than this because the velocity in the stream, though of the same order as the free fall velocity, must always be somewhat less because of the outward momentum of the laterally entering material continually being added to it. We thus have that

time of contraction of a segment
$$< \tfrac{1}{4}\pi R^{\frac{3}{2}}(GM)^{-\frac{1}{2}} < \tfrac{1}{4}\pi \sqrt{2}\,R^{\frac{3}{2}}(GM)^{-\frac{1}{2}} \tag{22}$$
$<$ time of fall to the sun of segment.

It follows that the internal forces are strong enough for the stream to divide into separate pieces before reaching the immediate neighbourhood of the sun. Moreover, the time of contraction $\frac{1}{8}\pi d(Gm)^{-\frac{1}{2}}$, for values of R comparable with GM/V^2, is of the same order of magnitude as the time required to build up the steady state, namely GM/V^3, which is the characteristic time associated with the accretion process. Thus the contraction into segments will not proceed so quickly as to cause the state of the system, especially in the outer part of the stream near the neutral point, to depart seriously at any time from the continuous steady state condition assumed in investigating the accretion process.

It is obviously an essential point of the theory that a comet can hold together by self-gravitation when it first forms at great distance. The presence of the factor $1/R^3$ in the sun's differential effect is clearly the feature that renders this possible despite the small masses and large sizes of comets.

Number of comets likely to form

The total amount of matter captured by the sun during its passage through a cloud can readily be estimated for any particular size of cloud. For example, if the cloud were 1 parsec in depth, or about 3×10^{18} cm., the time of passage of the sun through it would be about $10^6 v^{-1}$ years (where v is the velocity expressed in kilometres per second, as usual), and since material is captured at a total rate $2 \cdot 5\pi \rho G^2 M^2/v^3$ (for $\alpha = 1\cdot 25$), the total amount failing to escape would be at most about

$$4 \times 10^{26} v^{-4} \text{ g.}$$

for $\rho = 10^{-25}$ g. cm.$^{-3}$.

But even if the upper limit of mass of a comet, given by (21), were adopted as an average mass, this would be enough material for about $5 \times 10^5 v^5$ comets. This figure will tend to be an underestimate because of the probability that most of the comets formed may have smaller mass than the maximum. On the other hand, if only a small fraction, say only a few per cent, of all these actually avoid falling directly into the sun and become comets in orbital motion round the sun (as we

shall see shortly is likely to be the case for quite a high proportion), there would nevertheless be enough material captured
for the formation of some tens of thousands of comets during a
single passage through a cloud.

During its recent lifetime the sun may well have passed
through numerous clouds at different times with various directions of relative velocity, and accordingly it may legitimately
be supposed that the existing comets are the outcome of several
encounters during a comparatively recent stage of the sun's
history, possibly the last few hundred million years or even less.

An interesting feature of the accretion process of formation
of comets is that it could proceed without anything of it being
observable on Earth, always supposing the original dust cloud
itself not too extensive to involve obscuration of moderately
near stars. For the accretion stream is in effect itself a line of
comets, equivalent probably to a few thousand in number, and
no comet could be observed at several hundred astronomical
units—the record distance is only a little more than 10 a.u.
Thus the stream must remain unobservable except possibly in
the event that it was so situated that the Earth should pass
almost exactly across it so that the line of sight was roughly
directly along its length, in which case it might just possibly
be barely visible.

Why comets do not fall directly into the sun

In the idealized circumstances to which the solution of the
accretion problem accurately applies, the process clearly could
not result in the primitive aggregations of dust particles having
orbital motion round the sun. For the accretion axis along
which the stream moves is directed exactly through the centre
of force, and the net angular momentum of the accumulations
at the axis must therefore always be zero, by symmetry. Also
we have seen (Table II) that even at great distances from the
sun the width of the stream will itself only be at most of about
the same order of magnitude as the radius of the sun (about
7×10^{10} cm.), so there would be no possibility through this
feature either of any of the aggregations within the stream

96

managing to avoid striking the sun's surface, if the sun were the only centre of force within the system. But in the actual solar system there are known influences that prevent any difficulty arising. In the first place the presence of the planets and in particular Jupiter, which lies fairly near the sun at about 5·2 a.u. as compared with the distance of the neutral point at several hundred astronomical units, displaces the effective centre of attraction to a point that actually lies just outside the sun, and if the influence of Saturn and the other great planets is allowed for also the displacement may at times be even greater. It can readily be shown that any gravitating system attracts at great distances (compared with its own size) just as though all its mass lay at the centre of gravity of the system. This conclusion follows immediately from MacCullagh's well-known expansion formula for the potential of a mass system,* but we can demonstrate it directly for the sun and Jupiter in the following way, just as is done for the Earth-moon system in the lunar theory.

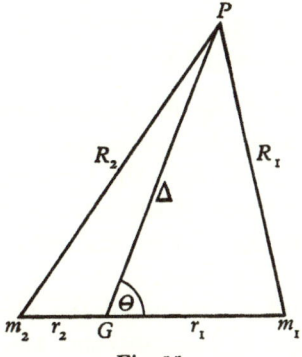

Fig. 11.

Suppose m_1 denotes Jupiter and his mass, and m_2 the sun, and the point G their centre of mass. Then if r is their distance apart, and we write $r_1 = m_1 G$, $r_2 = m_2 G$, we have

$$m_1 : m_2 : m_1 + m_2 = r_2 : r_1 : r,$$

* $(GM/r) + (G/2r^3)(A + B + C - 3I) + \ldots$ in the usual notation. See Jeffreys and Jeffreys, *Methods of Mathematical Physics*, p. 512.

7

by the centre of mass property. Now if we suppose P to be a point at great distance Δ from G compared with either r_1 or r_2, so that $\Delta \gg r$, then the potential at P can be developed as an expansion in powers of r/Δ thus, in the notation of Fig. 11:

$$
\begin{aligned}
\frac{m_1}{R_1} + \frac{m_2}{R_2} &= m_1(\Delta^2 + r_1^2 - 2\Delta r_1 \cos\theta)^{-\frac{1}{2}} + m_2(\Delta^2 + r_2^2 + 2\Delta r_2 \cos\theta)^{-\frac{1}{2}} \\
&= \frac{m_1}{\Delta}\left(1 + \frac{r_1}{\Delta}\cos\theta + \frac{r_1^2}{\Delta^2}\cdot\frac{3\cos^2\theta - 1}{2} + \ldots\right) \\
&\quad + \frac{m_2}{\Delta}\left(1 - \frac{r_2}{\Delta}\cos\theta + \frac{r_2^2}{\Delta^2}\cdot\frac{3\cos^2\theta - 1}{2} - \ldots\right) \\
&= \frac{m_1 + m_2}{\Delta} + \frac{m_1 m_2}{m_1 + m_2}\cdot\frac{r^2}{\Delta^3}\cdot\frac{3\cos^2\theta - 1}{2} + \ldots.
\end{aligned}
\tag{23}
$$

The first term is simply the potential that the total mass would have if concentrated at G, while no term in r/Δ^2 appears because G is the centre of mass. To estimate the size of the third order term compared with the first, their ratio is

$$
\frac{m_1 m_2}{(m_1 + m_2)^2}\cdot\frac{r^2}{\Delta^2}\cdot\frac{3\cos^2\theta - 1}{2},
$$

and we have $m_1/m_2 = 1/1047$ for Jupiter and the sun. Now at a part of the stream near the neutral point, where $\Delta \doteq 500$ a.u., we may take $r/\Delta = 1/100$. Also the term $\frac{1}{2}(3\cos^2\theta - 1)$ cannot exceed unity. So the value of the ratio is about 10^{-7} at most, and it follows that the force at great distance is almost exactly towards the centre of the mass.

A similar result follows when all the planets are included, the centre of attraction at great distance then being practically the centre of mass of the whole solar system. This is represented in the corresponding expansion by a first term $\Sigma m/\Delta$. The term of next higher order again disappears when Δ is measured from the centre of mass, and the next non-vanishing term, though more complicated in form, is again of order 10^{-7} for points at distances comparable with that of the neutral point.

Now the centre of gravity of the sun and Jupiter together lies at about $1 \cdot 07$ solar radii from the sun's centre, and so is always just outside the sun's surface. The terrestrial planets produce negligible displacement because of both their small masses and small heliocentric distances, but the other three great planets, Saturn, Uranus, and Neptune, when situated in about the same longitude as seen from the sun, so that they are more or less in line on the same side of the sun, can displace the centre by about 1·10 solar radii. It follows, therefore, that, according to the positions of the four outer planets, the centre of mass of the solar system and with it the effective centre of attraction, can lie at a distance from the centre of the sun ranging from zero up to nearly 2·20 solar radii. It is also evident that during any long interval, such as that taken by the sun to travel through a dust cloud, in which the relative positions of the planets, particularly Jupiter and Saturn, take almost all configurations, that also more or less repeat themselves, the proportion of time during which G lies outside the sun will be at least comparable with the time it lies inside.

Thus during such a part of the time of passage through a dust cloud, aggregations of primitive comets in the inward flowing stream will fall, while at great distances, towards a point lying outside the sun, and in this way will acquire angular momentum about the centre of the sun. The effect is due of course ultimately to the transverse attraction of the planets on the comet. As the distance decreases and a comet comes within the orbits of the outer planets, their effect diminishes and the motion is largely as if the sun alone acted. Thus at least a proportion of the comets would attain sufficient angular momentum to enable them to sweep round the sun clear of its surface, even allowing for the comet having finite size itself. But if only a few per cent of all comets forming in the stream escaped direct absorption into the sun by this means, though it is not suggested that the proportion would necessarily be very small, the total estimated number given by (23) shows that nevertheless an average cloud might easily produce several thousand comets.

The detailed investigation of the precise path near the sun that a comet would take after falling from a great distance under the combined influence of the sun and planets would constitute a major computational problem in celestial mechanics of a kind as yet nowhere tackled and is not attempted here. But such a study would obviously be of the greatest interest in relation to the present problem, for in the absence of its solution we can make only a general decision on what the nature of the motion is likely to be. With this reservation, we may conclude that the presence of the planets may result in a comet becoming endowed with sufficient angular momentum per unit mass to enable it to avoid striking the sun's surface at its first perihelion passage. Moreover, it seems probable that at each subsequent return to perihelion the action of the planets will again transfer to the comet a comparable amount of angular momentum, but always in a more or less random direction compared with that during previous perihelion returns and, dependent on the prevailing configuration of the planets, which will not bear any special relation to the original inward motion of the comet and so will be unrelated from one return of the comet to the next. In this way the perihelion distance would be expected in some cases to undergo progressive increase just as random errors are subject to progressive increase, though in the early stages it is entirely possible that for some comets the effect would be to reduce the distance again and perhaps even cause the comet to fall into the sun. But it is only necessary for the purposes of the theory that a proportion of comets escape such capture since, as we have seen, the total number of potential comets is very large.

Negligible effect of planetary perturbations on the accretion process

We can at this point consider the possible effect that the planets might have in deviating the dust particles from strictly hyperbolic orbits in their motion towards the accretion stream and thereby causing them to fail to meet the accretion axis. So far their motion has been considered as taking place solely under the simple inverse square attraction towards the centre of mass of the solar system as a whole. This force is represented by the first term of (23), but in addition there will be certain far smaller forces arising from the various terms due to the different planets similar to the second term of (23), and these at times would operate to deviate the particle out of the plane of the simple hyperbolic motion that accurately intersects the accretion axis. Any additional small disturbances within this plane will not affect the question, and it is necessary therefore to examine only the magnitude of effects perpendicular to it.

It is clear that the relevant deviating force will be greatest for particles whose motion lies (effectively) in a plane perpendicular to the plane of motion of the planets about the sun, as can in fact be rigorously shown from the expression for the force function given by the second term of (23). It follows that the greatest value of the sideways force, f say, due to any particular planet m_1 is of amount

$$f = \frac{3}{2} \cdot \frac{Gm_1 m_2}{m_1 + m_2} \cdot \frac{r^2}{\Delta^4} \doteqdot \tfrac{3}{2} Gm_1 \frac{r^2}{\Delta^4}$$

the maximum value having been taken of the derivative of the term in θ, which is what is involved in the required component of the force. In this expression G is the constant of gravitation, m_1 the mass of the planet, m_2 the mass of the sun, and r the radius of the orbit of the planet, while Δ is the distance of the dust particle from the centre of gravity of the two bodies. From the known values for the various planets, the product $m_1 r^2$ is found to be greatest for Neptune, for which it is about twice the corresponding value for either Jupiter or Saturn, and about two and a half times that for Uranus. The motion of the planet on either side the plane of undisturbed hyperbolic motion is periodic and the greatest deviation that may result during the time the planet lies to one side of this plane is easily shown to be $\tfrac{1}{4} f t^2$, where t is the time spent by the planet on one side of the plane, namely half its orbital period.

For Neptune this period is about 164 years, so that $t = 2 \cdot 6 \times 10^9$ secs. while $r = 30$ a.u., and if we take $\Delta = 1000$ a.u., it is readily found that the *maximum* possible deviation comes to about 7×10^4 cm. Allowance for the action of other planets

101

could at most increase this by a factor of about 2·4 leading to a maximum deviation of less than 2×10^5 cm. This distance is far less than the width of the accretion stream, which we have seen would probably be of order 10^{10} cm., or the average dimensions of long-period comets.

The maximum deviation would become comparable with the width of the stream only for values of Δ of about 50 a.u. or less, so that planetary perturbations will not prevent any particle from being absorbed into the stream except possibly at comparatively small distance from the sun. Even so these calculations refer to the maximum possible deviation, supposing all the planetary forces to act always in the same direction. For a high proportion of the particles and for most of the time these forces will not only tend to cancel each other out but also have less than the maximum value. Accordingly it may be concluded that although planetary perturbation might somewhat hinder the early stages of building up the accretion stream by causing some of the particles to fail to intersect the axis, once the stream had built up, the deviations would be negligible compared with its width.

The periods and eccentricities of the initial orbits

It is clearly indicated by the theory that, if no earlier strong perturbative force has intervened, all comets in their initial sweeps round the sun will pass close to its surface; that is, their perihelion distances will be at most a few solar radii. Where the rest of the initial inward orbit is concerned, that part which is practically the inward flowing stream, to a hypothetical cosmical observer situated at very great distance but assumed able to see the path of the comet, would appear to be just a straight line directly towards the sun (or rather the centre of mass of the whole system) and originating at a distance of the order of GM/V^2 from it. Thus this part may be thought of as a very narrow elongated ellipse having minor axis ($2b$) negligible compared with the major axis ($2a$), and therefore having eccentricity close to unity, since $b = a \sqrt{(1 - e^2)}$. Accordingly these initial orbits would appear practically parabolic.

Where the period is concerned, for a comet arriving from somewhere near the neutral point we shall have, at any rate in order of magnitude, for the semi-major axis

$$a = \tfrac{5}{8}GM/V^2 = 5\cdot55 \times 10^2 v^{-2} \text{ a.u.}$$

By Kepler's law, the period in this orbit expressed in years is simply $a^{\frac{3}{2}}$ or about $1\cdot3 \times 10^4 v^{-3}$ years. This then gives a measure of the period that would be obtained if observations of the comet along the whole inward orbit were possible. In practice, however, observations can only be made at most for that part of the orbit within a few astronomical units of the sun on either side of perihelion, and there is no reason to suppose that the period derived by fitting an elongated ellipse (if this can be done at all) to the perihelion part, as closely as possible, will give a period agreeing with the preceding value, except perhaps in order of magnitude. Nor is it to be expected that in the next sweep round the sun, comprised of the first outward motion and the second inward motion, the time taken will be the same as either of the two former periods, for the slightest perturbation of a nearly parabolic orbit, especially when near the sun, will cause it to undergo very large changes elsewhere. All we can be fairly sure of is that so long as the comet does not pass exceptionally close to a planet the order of magnitude of the period is not likely to fall greatly below the original value though it might easily exceed it considerably at a subsequent revolution, since the binding energy of a nearly parabolic orbit, viz. $GM/2a$, is so small.

Of special interest in this connexion is an investigation by H. N. Russell, who showed in 1920 that for long-period comets on the average, the quantity $1/a$, which in effect measures the total dynamical energy (potential plus kinetic) of the comet, is likely to undergo random changes by amounts of order 6×10^{-4} (a is here measured in astronomical units) during the comparatively small part of its orbit at which the comet moves at distances from the sun comparable with those of the great planets, in particular Jupiter. Since the period P in years is, by Kepler's law, simply $a^{\frac{3}{2}}$, it follows that variations in the two are connected by

$$\delta P = \tfrac{3}{2} P^{\frac{1}{3}} \delta(1/a) = 9 \times 10^{-4} P^{\frac{1}{3}} \tag{24}$$

for Russell's calculated *average* value of $\delta(1/a)$.

Thus for a comet for which the distant portion of the inward orbit was such that a period of 10^4 years would be deduced from it (that is, the total time of inward motion would be 5000 years), the period would stand an even chance of being changed through the effect of the planets to a new value within the range $10,000 \pm 4000$ years, approximately, while there is a smaller but not negligible chance of it undergoing much greater deviation. For example, if it should pass sufficiently near Jupiter, as might happen exceptionally, it could be deflected into a short-period orbit of a few years period, or it could even be deflected into a hyperbolic orbit, with theoretically infinite period, and be lost from the solar system. We may thus conclude that because of the actions of the planets the periods of comets, determined originally by the circumstances of the accretion process which settle the range $\alpha GM/V^2$, will after a few revolutions become widely dispersed from the initial values and soon show little trace of the first distribution.

These considerations make it quite plain that little reliance can be placed, from the point of view of cosmogonical significance, on any arguments concerning the present periods except perhaps the general order of magnitude of the periods of comets of very small perihelion distances which may not have been seriously disturbed, and even this is doubtful. An increase of V by a factor of 2 would reduce the distance of the neutral point by a factor of one-quarter, and accordingly shorten the initial periods by one-eighth, as compared with a corresponding system generated from a cloud of smaller V. But the disturbing action of the planets would now operate to spread these periods in both directions, and thereby would soon bring about comets of far longer periods than would be suggested simply by the length of the inward accretion stream.

The point is further illustrated by a table of values compiled by Strömgren of all those comets found to have definitely hyperbolic motion near the sun (that is, a value of e greater than unity by an amount not ascribable to admissible errors of observation) and for which the earlier part of the orbit has been found by computation carried backwards in time making allowance for planetary attraction. In view of its special relevance to hypotheses on the origin of comets we give here some details of this table.

TABLE IV

Comet	Time t, from perihelion (years)	$1/a$ from orbit near perihelion (a in a.u.)	$1/a$ from earlier orbit beyond those of planets	$\delta(1/a) \times 10^4$
1853 III	15	−0·000819	+0·000083	9·02
1863 VI	18	−0·000495	+0·000017	5·12
1882 II	14	+0·011896	+0·012149	2·53
1886 I	13	−0·000694	−0·000007	6·87
1886 II	11	−0·000477	+0·000317	7·94
1886 IX	12	−0·000576	+0·000063	6·39
1889 I	19	−0·000692	+0·000042	7·34
1890 II	12	−0·000215	+0·000072	2·87
1897 I	17	−0·000872	+0·000040	9·12
1898 VIII	12	−0·000607	−0·000016	5·91
1902 III	18	+0·000081	+0·000005	0·76
1904 I	14	−0·000504	+0·000216	7·20
1905 VI	10	−0·000142	+0·000621	7·63
1907 I	17	−0·000499	+0·000025	5·24
1908 III	20	−0·000732	+0·000158	8·90
1910 I	12	+0·000214	+0·003302	30·88
1914 V	13	−0·000146	+0·000012	1·58
1922 II	28 (52 a.u.)	−0·000381	+0·000004	3·85
1925 I	21	−0·000566	+0·000054	6·20
1925 VII	9	−0·000273	+0·000115	3·88
1932 VI	5 (16 a.u.)	−0·000595	+0·000044	6·39
1936 I	20	−0·000487	+0·000205	6·92
			Average value of $\delta(1/a) = 7\cdot2 \times 10^{-4}$.	

The heliocentric distances, corresponding to the time t from perihelion (of the second column), are given approximately by the formula $5\cdot6\ t^{\frac{2}{3}}$ a.u., and range from 16 a.u. for 1932 VI to 52 a.u. for 1922 II.

The orbit of 1914 V has also been computed forwards in time (from 1914) to a distance of about 39 a.u. by van Biesbroeck who finds an eventual value of $a^{-1} = +0\cdot000126$, which corresponds to an aphelion distance of only about one-tenth its preceding value, and implies a change in 'period' from one revolution to the next by a factor of about thirty.

From the last column it is seen that the average value of $\delta(1/a)$ comes to about 7×10^{-4}, in full confirmation of Russell's estimates, but it has to be remembered nevertheless that the distances from the sun which the calculations covered in several cases only slightly exceed the radius of Neptune's orbit, so that some very small further change before a 'final' value of $1/a$ would be found for the particular circuit of the sun concerned may be expected in such cases. The table brings out again the point referred to earlier of the impossibility of deciding accurately the period of any long-period comet from observations near perihelion. Of course for the present hyperbolic comets there would appear to be infinite period near perihelion, but if for example we consider the values of $1/a$ near perihelion for 1904 I and 1907 I, the table shows these to be almost the same, viz. $-0 \cdot 000504$ and $-0 \cdot 000499$. Yet the elliptic values at great distance for these two comets are $0 \cdot 000216$ and $0 \cdot 000025$ respectively, and these correspond to simple Keplerian periods of about 300,000 years in the one case and 8,000,000 years in the other. Moreover, there is no reason why at a later, or earlier return of such comets equally large changes in period, as are indicated by the difference in these two values, may be or have been produced, and also a future change to a hyperbolic orbit involving complete escape. No such hyperbolic orbit has yet been found because (with one exception, see Table IV footnote) the computations so far made have all been carried out *backwards* in time to give the inward path, and necessarily any observed comet formed by accretion by the sun will have started its inward motion from a finite distance. If such calculations were carried *forwards* in time for every long-period comet there is little doubt that instances would soon be found of actually observed comets destined to leave the solar system in slightly hyperbolic orbits.

Other disturbances of the orbits

The progressive increase of the least distance from the sun, q, as a result of the foregoing process cannot have the effect of transforming a long-period orbit into a short-period one during a single approach to the sun unless the change in $1/a$ is many times greater than the average value 6×10^{-4}. On the other hand if the orbit of a long-period comet happened to pass near that of Jupiter, there is a small but finite probability that a really close encounter would occur during which the attraction of the planet temporarily becomes comparable with or even exceeds that of the sun on the comet. In such circumstances the comet would generally receive angular momentum per unit mass comparable with that carried by the planet and so be deviated into an orbit with a semi-parameter, that is a value of $a(1-e^2)$ whose square root measures the angular momentum, comparable with the radius of Jupiter's orbit. Now since for the least distance

$$q = a(1-e) = \frac{a(1-e^2)}{1+e} > \tfrac{1}{2}a(1-e^2), \quad (\text{since } e < 1),$$

it follows that after such a deviation q will as a rule be much larger than for a sun-grazing orbit.

It appears to be generally agreed that it is by this process that the short-period comets, comparatively few in number, have come to move in their present paths. There is no reason to expect that the resulting paths will be even approximately parabolic, and the small inclinations and eccentricities of most of their orbits, notably 1925 II and Comet Oterma for which $e = 0.14$, are in full accord with the hypothesis of deflexion by Jupiter. All this has of course been a settled theoretical matter for many years, quite independently of any theory of the ultimate origin of comets. The process is often referred to as 'capture' by Jupiter, but it is important to understand that on the present theory of formation of comets, the comets concerned in the encounters with the planet have already been captured by the sun through the dissipation of energy involved in the accretion process, though as a result becoming only comparatively weakly bound in long-period motion. The later 'capture' by Jupiter is a subsequent event that may not happen for many revolutions, and indeed for most comets, may never happen at all. In such encounters, however, energy is conserved, and the great changes of orbit occur in a three-body motion in which essentially the sun, Jupiter, and the comet are concerned.

The theory of such three-body encounters has been extensively studied by H. A. Newton (1893), and by H. N. Russell (1920) in the work already referred to, but their researches are too technical and extensive for more than a general mention of the conclusions to be included here. Their main results may be briefly stated as follows.

It is possible for an originally long-period almost parabolic orbit to be transformed by a single encounter with any one of the major planets into an elliptic orbit of moderate or small eccentricity and with a period shorter than that of the planet. Such large changes will occur only when the encounter is very close and therefore they are correspondingly rare, so that for every instance of a large change there will be a considerably larger number of cases of minor perturbations. Of comets captured in this way, by far the majority will move subsequently in direct orbits with fairly small inclinations, as in fact the short-period comets are nearly all observed to do. Moreover, the perturbed orbits will continue to pass near the orbit of the planet originally responsible for the capture. From this consideration Russell has shown that practically all the known short-period comets are to be associated with Jupiter.

The importance of all this to the present hypothesis of the formation of comets is that it makes it unnecessary for the theory to provide for comets moving initially in such orbits as the short-period comets exhibit. As we have seen, all comets will begin their existences in orbits approximating to a straight line directed towards the centre of the solar system, and the subsequent strong perturbative action of the planets on the small proportion that ever happen to move suitably, either at first or later through the gradual effect of perturbations, is an inevitable dynamical process.

There is, however, the further possibility of perturbations of long-period comets by passing stars when the comet is at great distance during the aphelion portion of its orbit. It has been seen that the initial distance to which a comet will recede may be of the order of several hundred astronomical units and that, through perturbations, this distance may be considerably increased for later aphelion excursions in the case of some comets. Now from the known density and motions of the stars in space in the neighbourhood of the sun, it can be estimated that in 10^9 years, a number approaching 100 stars will have passed within less than 10,000 astronomical units of the sun. A star passing at this distance would be certain to produce perturbations on a comet that happened to be suitably placed at great distance from the sun at the time, and at such a distance only a very small transverse force would be needed to impart considerable angular momentum about the sun to the comet compared with the extremely small amount per unit mass that a sun-grazing comet would have.

It is to be noticed, however, that a comet deflected in this way from a highly elongated path would still have to retain an eccentricity near to unity if it is later to become observable. For in the first place such a stellar perturbation would normally occur only when the comet was at great distance, so that the new orbit would necessarily still have large aphelion distance, that is

$$a(1+e) \gg 100 \text{ a.u.}$$

at the very least, while in order that the comet should become discoverable, its perihelion distance from the sun must certainly satisfy

$$a(1-e) < 5 \text{ a.u.}$$

If the sign of equality were adopted instead in both these conditions the resulting value of e would be an extreme lower limit for such comets, but this nevertheless exceeds 0·9. In most cases these conditions would probably be very strongly satisfied and the resulting eccentricity accordingly much closer to unity. In other words, near the sun the path of such comets would necessarily be found to be almost parabolic.

There are reasons for believing, however, that the present comets of the solar system are more recent developments than the planets, with the oldest perhaps at the very most a few hundred million years old, though of course by no means are they necessarily all of the same ages if they have resulted from passages through several clouds. If this is so, few passing stars may have come near enough to the sun to produce effective perturbations. Thus although it is possible that there are numerous comets moving with perihelion distances too large for observation of them to be made by present means, it seems probable that their total number may not necessarily be great, perhaps only a moderate addition to the entire number 200,000 estimated to be capable of becoming observable by present-day methods.

IV

THE FORMATION OF TAILS

The theory of the preceding chapter has shown how, in agreement with general ideas already arrived at by astronomers, comets consist of very large numbers of widely separated particles. With this picture of the internal structure of a comet, we come now to a simple consequence of the theory that throws considerable light on the whole question of the formation of tails. The difficulty has always been to understand why comets should more or less suddenly begin to emit material for tail formation and how they are able to go on repeating the cycle at each perihelion return more or less indefinitely.

It has been seen that as a result of gravitational forces within the accretion stream the cometary segments form initially at great distance from the sun and so give rise in the first place to long-period comets. Also, since by equation (19) of Chapter III the mass md depends on $R^{\frac{3}{2}}$, comets of greatest mass will tend to form at the greatest distance. Now as such a comet falls towards the sun its self-gravitation will remain unchanged, at any rate in order of magnitude, whereas the differential force on it due to the sun will increase like R^{-3}. Since R decreases from several hundred astronomical units down to a few solar radii as the comet moves inwards, the effect of the sun's action increases by a very large factor. For example, we might have initially $R = 500$ a.u. $= 10^5$ solar radii, and if the comet eventually passes within about 2 solar radii of the sun's centre, the differential force would be increased as compared with the comet's internal self-gravitation by a factor of about 10^{14}. This completely reverses the relative importance of the two effects; at great distance the internal gravitation predominates, though probably only by a moderate factor, whereas when close to the sun the internal gravitation is utterly negligible.

111

The same result holds good even though the comet is not a sun-grazer (i.e. $q \sim$ radius of the sun), and the point is of such importance to our argument that it may be worth while examining it more generally. If m_c denotes the mass of the comet, the force at distance r from it, in gravitational units, will be m_c/r^2. But if R is the distance of the comet from the sun, the differential force due to it at points distant r apart in the radial direction from the sun (for which the differential force is greatest for given displacement r) is $2Mr/R^3$. The sun's influence will therefore exceed that of the comet in the ratio

$$2\frac{M}{m_c} \cdot \frac{r^3}{R^3}, \tag{1}$$

and a few numerical instances show that this number is always large at any position at which a comet could be observed. Thus:

(i) *Encke's comet.* If we suppose the mass to be 10^{17} g., probably an overestimate, then it is readily found that at perihelion, where $R = 0.34$ a.u. and the observed radius $r = 2.4 \times 10^8$ cm., this ratio comes to 4×10^3 in round numbers. On the other hand, at aphelion, where $R = 4.1$ a.u. and $r \geqslant 2.5 \times 10^{10}$ cm., the ratio is about 9×10^6 at least.

This particular comet often appears of fan-shaped form and has on occasion exhibited great apparent contraction of radius at perihelion; the foregoing estimates for r, given by Chambers, can therefore only be regarded as correct in order of magnitude and applying to particles near the *observable* limit of the comet. For particles further out this ratio will be increased, and the sun's influence on them will be in even greater proportion to the attraction of the comet.

(ii) *Halley's comet.* If we suppose $m_c = 10^{18}$ g. and $r = 4.5 \times 10^{10}$ cm., then at perihelion ($R = 0.59$ a.u.) the ratio is found to be 5×10^8. At aphelion ($R = 35.31$ a.u.) no observations are possible, so adopting the same value for r will probably much underestimate it, but the ratio nevertheless comes to 3×10^3.

(iii) *Typical large long-period comet.* Such a comet might have $m_c = 10^{20}$ g. and $r = 10^{11}$ cm. In that case it is readily

PLATE XI

A. Brooks' Comet (1893 IV) 1893 October 21, showing remarkable irregularity in the tail at great distance from the head

B. Comet 1910 I (1910 January 27), showing curvature of tail

PLATE XII

HALLEY'S COMET 1910 MAY 7 (24-minute exposure),
showing curious tail structure

found that at R a.u. from the sun the ratio is about $10^7 R^{-3}$. This again shows the enormous extent to which the sun's action predominates. The internal effect only rises to comparable importance at about 200 a.u. from the sun. But as most long-period comets have periods considerably in excess of 1000 years, this means that when they are at great distances the internal gravitation may exceed the sun's differential field. Thus at 1000 a.u. the ratio would be 10^{-2}, implying that the self-gravitation would there exceed the sun's (differential) influence one-hundredfold.

The meaning of these results is clear, for if a comet consists of an aggregate of separate particles, their mutual forces must be entirely negligible, and hence *each particle of the comet must describe a separate independent orbit round the sun*. If the main proportion of the mass of the comet resides in a nucleus, the value m_c/r^2 for the force will closely apply everywhere, but if the mass were distributed more uniformly through the comet, the actual force at internal points would be less than m_c/r^2, and the result will still hold good.

This is perhaps a suitable point to refer to a mistaken impression that seems to have been held by a number of writers on comets concerning what they term the 'tidal influence' of the sun on a comet. As already mentioned (Chapter II, p. 49) the condition for stability of form of a *liquid* or *gaseous* body of mass m and radius r, against tidal disruption by the sun when at distance R from it, is necessarily of the form $Mr^3/m_c R^3 \ll 1$, and this condition is far from being satisfied especially at perihelion if applied to comets, as we have seen. But it is not permissible on such a basis to maintain that comets must therefore be tidally unstable and likely to disrupt into fragments at perihelion, because there is no reason to suppose that a comet can be regarded as a fluid system to which hydrodynamic theory applies. The idea of pressure, implied in the condition for the free surface of a body to be an equipotential, obviously does not apply to a cloud of particles moving independently of each other. For example, if we imagine two small dust particles pursuing exactly the same

8

orbit about the sun at a small distance apart, with negligible attraction between them, there is no question of any instability. Their distance apart will increase and decrease in proportion to the velocity, but no permanent change will occur.

Returning now to our main theme, let us accordingly regard a comet as an aggregate of a very large number of dust particles describing separate orbits but with each particle having precisely the same period. (The question of differences of period

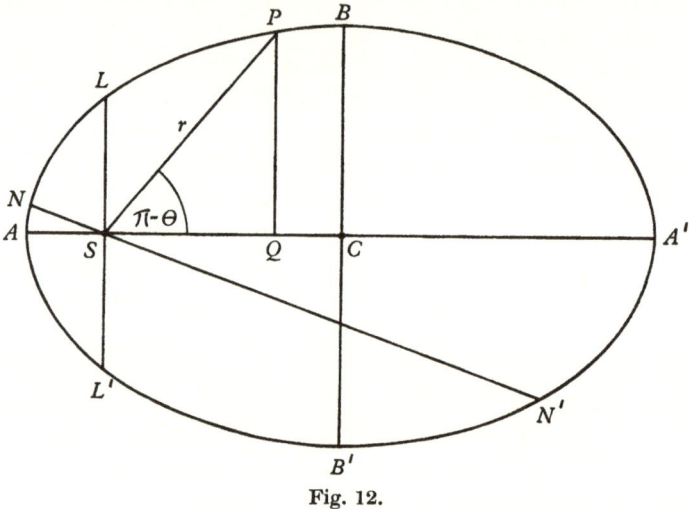

Fig. 12.

will be considered later.) This assumption must in any case hold to a high degree of approximation for a long-period comet, since at great distance from the sun the particles can hold together by self-gravitation and so they will all be approaching with practically the same velocity. The equation of the path of a particle situated at the centre of the comet may be written in polar form

$$\frac{a(1-e^2)}{r} = 1 + e \cos \theta \qquad (e < 1) \tag{2}$$

We can refer to this as the standard plane. As is well known, the period depends only on the major axis $2a$. If other particles of the comet have exactly the same period, each particular orbit will have exactly the same size of major axis and each will of course have the sun as focus. If Fig. 12 represents the standard orbit and ASA' the major axis, then it is clear that a possible adjacent orbit for any other particle of the comet can be obtained, without changing the eccentricity, by slightly altering the direction of the line ASA' without changing either its length or the relative position of S upon it. A possible orbit of the same period can also be got by a slight rotation of the standard orbit about AA'. Again, it is possible for the eccentricity e to be slightly changed, without change of a, and an adjacent orbit still obtained. Lastly, we notice that it would be possible for another particle to move in precisely the standard orbit but differ slightly in time of perihelion passage. All these changes are of course representable by slight changes in the standard orbital elements without change of a, and even if all occur together they will give an adjacent orbit provided they are small.

We see that these changes can be thought of as occurring in two ways. First of all within the standard orbital plane slight change can be made in e, and in the time of perihelion passage, and also a slight rotation can be considered of the standard orbit about a line through S perpendicular to the standard plane. But these changes will produce orbits all in the same plane as the standard orbit. Second, however, it is possible to rotate any of these orbits about any line through S, such as NSN' of Fig. 12, and obtain an adjacent orbit in a plane slightly inclined to the standard one. Moreover, part of such an orbit will always necessarily lie above the standard plane and part below it, *so that a particle describing such an orbit must cross through the standard plane twice during each revolution.*

Now there is no reason whatever suggesting that a comet is a flat distribution of particles confined to the standard plane, indeed, quite the contrary, for their extents perpendicular to the orbital plane, when they can be observed, appear to be always at least comparable with any other dimension. Accordingly, apart from just a few particles that may happen to move in the standard plane, every particle of a comet must

115

pass through this plane when at what we may term the nodes N, N' of its orbit. Since, however, the line NN' necessarily passes through S, it follows that every particle crosses through the standard plane during the time the comet as a whole is travelling the perihelion part of its orbit, that is, the part LAL' between the ends of the latus rectum.

By Kepler's equation, $nt = \phi - e \sin \phi$, relating the time from perihelion to the eccentric anomaly, it is readily found that the time spent by the comet in traversing the perihelion arc LAL' in terms of the total period P is

$$\frac{1}{\pi} [\cos^{-1} e - e \sqrt{(1 - e^2)}] P,$$

and if e is nearly 1, so that $1 - e$ is small, as is true for most comets, this interval can be expressed approximately as

$$\tfrac{3}{5}(1 - e)^{\frac{3}{2}} P,$$

and is then very small compared with P. For example, for Halley's comet, for which $P \doteqdot 75$ years and $e = 0 \cdot 9673$, this time is $0 \cdot 0034 \, P$, or about 3 months; while for some long-period comets, for which e is very close to unity, this part of the orbit may be described in a matter of a few days or hours, even though the total orbital period may be measured in hundreds, or thousands, of years. It thus appears that in a time of this order the whole comet must 'turn itself inside out', as it were, particles above the plane on the aphelion side of the orbit now move below it, and vice versa.

There is, however, an additional feature strongly tending to shorten still further the time during which this crossing-over occurs. For if, as would be expected, the nodes N' for the various particles are distributed more or less uniformly round the orbit, then the corresponding nodes N in the portion LAL' will obviously tend to be strongly concentrated near A, since the eccentricity is nearly unity. This will happen simply because of the elongated form of the ellipse and the fact that a far greater proportion of it lies on the aphelion side. In the same way, if the nodes N' were distributed round the standard

116

orbit at equal time intervals apart, an exactly similar result will hold because the comet moves so much more slowly at aphelion. If n is the mean angular motion, the angular distribution of the points N can easily be found by expressing $\dfrac{n}{2\pi}\,dt$ as a distribution function in terms of θ. This leads without difficulty to

$$\frac{n\,dt}{2\pi} = f(\theta)\frac{d\theta}{\pi} = \frac{(1-e^2)^{\frac{3}{2}}\,(1+e^2)}{(1-e^2\cos^2\theta)^2}\cdot\frac{d\theta}{\pi}, \qquad \left(-\frac{\pi}{2} < \theta < \frac{\pi}{2}\right), \qquad (3)$$

where θ is measured from perihelion. From this function, for Encke's comet and Halley's comet, the following values have been calculated of the proportion between $-\theta$ and $+\theta$ of the total integrated values between $-\dfrac{\pi}{2}$ and $+\dfrac{\pi}{2}$.

TABLE V

Encke $e = 0\cdot846$		Halley $e = 0\cdot967$	
θ	Proportion within $\pm\theta$	θ	Proportion within $\pm\theta$
10°	0·31	10°	0·65
20°	0·55	20°	0·88
30°	0·70	30°	0·95
40°	0·80	40°	0·98
60°	0·90	60°	0·99

In both cases the figures show the strong concentration towards perihelion of the nodes N, and a comparison of the two sets of values show how rapidly the degree of concentration increases as e increases towards unity. It is clear that these effects arise from the closeness of e to unity and will be almost independent of any particular assumption as to the precise distribution of the nodes.

Release of material in form suitable for the production of tails

This brings us to a simple consequence of the theory that immediately explains the source of material going to form the tails of comets. Hitherto the difficulty has always been to understand why certain comets should more or less suddenly begin to emit material for tail formation and how they are able to continue repeating this process at revolution after revolution.

As already mentioned in earlier chapters the actual extents of comets, in the direction along their orbits and in directions perpendicular to this, may much exceed the observed extents of their comas. The following values are given by Chambers as instances of dimensions of comas:

TABLE VI

	'Radius' of coma (cm.)
Holmes' comet (1892 III)	$1 \cdot 2 \times 10^{11}$
Comet 1811 I	$9 \cdot 2 \times 10^{10}$
Halley	$4 \cdot 5 \times 10^{10}$
Encke (greatest observed)	$2 \cdot 5 \times 10^{10}$
(least observed)	$2 \cdot 5 \times 10^{8}$
Comet 1849 II	$6 \cdot 5 \times 10^{9}$
Comet 1847 I	$3 \cdot 2 \times 10^{9}$
Comet 1847 V	$2 \cdot 0 \times 10^{9}$

From what has been said earlier, it is not to be supposed that these comets are necessarily spherical in form, and the term 'radius' can refer only to the average estimated radial extent of the observable portion transverse to the line of sight. The greatest widths of meteor streams also suggest that comets may have far greater extents than can be directly detected observationally. Showers lasting many days (about twenty days appears to be the longest) are known and in one day the Earth describes 3×10^{11} cm. along its orbit. Thus the actual 'radius' of a comet may well exceed the observed 'radius' very considerably, and even though the space density of particles in the unobservable region must be lower than in the coma, the total mass present there may nevertheless be comparable with that within the coma. The transverse dimension of a comet perpendicular to the orbital plane may therefore easily exceed the observed size by a factor of at least two. Referring now to

Fig. 13, the line BSB' represents the standard orbital plane viewed end-on, and C and C' represent two particles of the comet moving in orbits whose planes are slightly inclined to the standard plane, on opposite sides of it. We have seen that the nodes N must be concentrated closely about perihelion A, so for simplicity we can regard them as being at A. The paths of C and C' will then lie in the planes CSC_1 and $C'SC_1'$, as shown, and the point of intersection of the paths will be at A, which in the projected figure

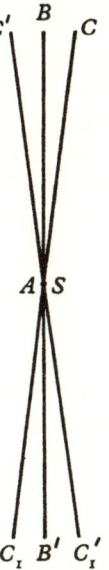

Fig. 13. End-on view of two orbital planes CSC_1 and $C'SC_1'$ inclined at small angle i to the standard plane BSB'.

coincides with S. If the inclinations of these planes to the standard plane are i and $-i$, where i is of course small, the relative velocity of the particles at A will be

$2i \times$ velocity of comet at perihelion.

The effective angular size of a comet on this basis will be the ratio of its diameter ($2\delta H$) perpendicular to the standard plane to the minor axis of the orbit. The following are estimates of the various quantities concerned for a number of comets:

TABLE VII

Comet	Estimated radius δH (cm.)	Angular size $2i$	Perihelion velocity of comet (cm./sec.)	Relative velocity of particles (cm./sec.)
Encke	5×10^{10}	6×10^{-3}	7×10^{6}	4×10^{4}
Halley	10^{11}	3×10^{-3}	6×10^{6}	2×10^{4}
Holmes	3×10^{11}	9×10^{-3}	$2 \cdot 5 \times 10^{6}$	2×10^{4}
Typical long-period comet	$\sim 2 \times 10^{11}$	$\sim 2 \times 10^{-3}$	$\sim 3 \times 10^{7}$	$\sim 6 \times 10^{4}$

We see therefore that in all cases the relative speeds at the standard plane range from about 0·2 to 0·5 kilometres per second.

Now since all the particles of the comet will be attempting to cross through the standard plane within a comparatively short interval equally spaced on the two sides of perihelion, it follows that we have here again a situation where collisions will be highly likely to occur for a proportion of the particles. Speeds of 0·2 km. sec.$^{-1}$, small as they seem astronomically, are nevertheless quite large compared by ordinary terrestrial standards; it would require free fall under gravity from a height of about 2 km. to produce it. There is no doubt that the impulses on particles resulting from such collisions would produce internal stresses far exceeding the strength of ordinary materials. Thus one of the main effects of the process would be to pulverize some of the particles of the comet and produce much smaller particles. The speeds involved here are lower by a factor of order 10 than those considered (Chapter III, p. 82) in the accretion process, so that there could be little possibility in this present case of more than a small proportion of the material being vaporized.

In arriving at the present estimates of the collisional velocities we have considered particles coming from extreme distances on either side of the standard plane, but the break-up of particles would still proceed even if the speeds were far smaller. For instance, to take a homely example, when a load of coal, or rubble, is tipped out from a height of a few feet above the ground, the collisional speeds involved are of order 3×10^2 cm. sec.$^{-1}$—about a hundredth of the above values—but as everyone knows, much of the coal, or stones, gets broken into smaller pieces and quite a cloud of dust is produced. So the estimated relative speeds within a comet can far exceed the values necessary for disruption of its particles. The proportions of different sizes produced would probably be roughly in accordance with a Poisson distribution, but the parameter would be difficult if not impossible to settle theoretically. There is little doubt, however, that a practically continuous distribution of sizes would result, and therefore that in any particular range of size a small but finite proportion of particles will be found.

Now it has long been understood that the tails of comets could be produced by the pressure of sunlight on dust particles with linear dimensions of the same order as the wave-length of the light incident upon them. It has been shown by Schwarzschild, Nicholson, Proudman, and others, that for a considerable range of sizes of this order, namely about 10^{-5} cm., the outward repulsive force attains to about twenty times the gravitational attraction of the sun. Thus, during the part of its orbit nearest the sun, through these simple dynamical causes, a comet must automatically release a proportion of dust particles of smaller sizes than its existing particles, and as soon as this occurs those in the critical range of size will be 'blown' away by radiation pressure to form the tail.

Proportion of material involved in tail-producing collisions

A first estimate of the probability that any particular particle crossing through the standard plane undergoes a collision can be made from consideration of the brightness of comets when at great distance from the sun. At this stage most comets are in a fairly quiescent state and shine mainly by reflected sunlight. According to H. N. Russell, from photometric observations of Halley's comet, the brightness is such that at this time only about 1/300,000 of the whole area within its apparent boundary is actually occupied by material, the comet being regarded for the moment as a two-dimensional object projected on the background of the sky. Thus the probability of collision of a particle similar in size to those of the comet and moving through it, making some allowance for its size, would be of order not less than 10^{-5}. If now only this proportion of the total mass, say 10^{18} g., were involved in collisions through the present cause, it would nevertheless mean a quantity 10^{13} g. On the other hand, there is little doubt that the occurrence of collisions near the standard plane will tend to cause material to accumulate there and so lead to a higher probability of further collisions, as in the accretion process itself, though now

121

instead of being a line the collision zone is ideally a plane. Accordingly this estimate may be regarded as a lower limit to the proportion of the comet's mass involved in collisions. Of course only a small proportion of the material will be transformed into particles of the appropriate size, but there is little doubt that it would nevertheless be adequate to produce all the required tail material, during any one sweep round the sun.

The process makes clear how a comet can go on producing innumerable tails, for even in the extremely unlikely possibility of the whole amount of 10^{13} g. being lost from the comet during a single perihelion passage, the comet could yet undergo 10^4 such passages before losing one-tenth of its total mass. The shortest known period is that of 3·3 years for Encke's comet, and the total time involved would therefore be at least 33,000 years, and no observations extend over anything like such a range. Even for Halley's comet, with its period of about 75 years, records go back only a little more than 2000 years, but 10^5 revolutions of the comet (supposing its period to remain as short as the present value, which would be highly doubtful) would take 750,000 years. Accordingly even for short-period comets the process makes clear that material for tail-formation may continue to be supplied more or less indefinitely so long, practically, as the comet itself continues to exist.

Comets without tails

On the present theory there is no difficulty in perceiving what factors are concerned in tail formation, and what types of comets might be expected to show little or no tails. There are evidently four main factors involved, two of which are intrinsic to the comet itself, and two refer to the orbit. They are the following:

(i) *The dimension of the comet perpendicular to the standard plane.* The greater this is, the greater in general will be the collisional velocities. Some diminution might gradually result from the loss of transverse motion due to inelastic collisions if there are no other causes tending to expand the comet.

(ii) *The existence within the comet of a sufficient density of dust particles greater than the critical size,* so that collisions can produce particles suitable for tail-formation. (If a comet consisted mainly of particles smaller than the critical size, their further break-up could not produce any supply of particles in the critical range.)

(iii) *The smaller the value of q, that is a(1 − e), the greater the velocity of the comet near perihelion and the greater the velocities of collision within the comet, other things being equal.*

(iv) *The closer the eccentricity is to unity the shorter the time taken to describe the perihelion portion of the orbit between the ends of the latus rectum in comparison with the whole orbital period.*

Now taking these factors in turn, it is clear that collisions at the standard plane will tend to diminish the extent of the comet above and below the plane, but on the other hand external perturbations might offset this. Where the particle size is concerned, it is entirely possible that a gradual decay of the comet's tail-forming ability would result, but as there is no reason to expect that more than a small fraction, say less than $1/1000$, of the whole mass is involved in collisions at each revolution, it is doubtful if this would lead to any detectable decrease in activity over the period of time for which observations have been possible. The least distance from the sun clearly affects the degree of violence of the collisions, and we would therefore in general expect conditions for tail-formation to be more favourable the smaller q is.

Where the eccentricity is concerned, however, so long as this is near unity, the activity on the perihelion side of the orbit will take place during a very short interval. As each particle necessarily passes through the standard plane twice, collisions will always be occurring at all other parts of the orbit too. But there are several influences tending to make any resulting effects here far less intense than for the perihelion side of the orbit. First, the velocity of collisions will be less because the orbital velocity of the comet is less—far less if e is close to unity. Second, the time spent on the aphelion side of the orbit for e near to 1, is so much longer that the collision rate will be proportionately reduced and in a given interval so few particles will collide that a negligible amount of fine dust will be produced. For instance, for Halley's comet the ratio of the time spent on the aphelion side of the orbit to that on the perihelion side is about 300 to 1, while the ratio of velocities at perihelion and aphelion is about 80 to 1. These figures make plain that tail-forming activity near aphelion will be negligible in comparison with that near perihelion. But in addition to these factors there is the important consideration that no comet of high orbital eccentricity can be observed at aphelion, or indeed anywhere near aphelion.

On the other hand there are periodic comets observable, in suitable circumstances, at a large part of their orbits, and in two cases all round the orbits. For this latter to be possible, however, the eccentricity must be much less than unity, and then the intervals of time spent on the perihelion and aphelion sides of the orbit are always roughly comparable. For these comets, Schwassmann-Wachmann and Oterma, for both of which e is approximately 0·14, the ratio of the times in the two portions of the orbit is about 2 to 3, while the ratio of speeds is about 1·15 to 1. Thus there would be no reason to expect any particular part of the orbit of such comets to possess any marked difference from any other part, as far as internal activity due to the present cause is concerned. The first of these two comets undergoes large irregular fluctuations at, it seems, randomly spaced parts of its orbit, and if these are due to causes of the present kind it is possible that they may arise from some irregularity of form of the comet. Irregularly distributed patches of dust associated with the comet, but more or less detached from it, would cross through the main part of the comet twice per revolution, and could thereby cause temporary activity. This is yet only a tentative suggestion but it is one that could be investigated by noting the precise positions of the comet when activity occurs and testing whether any of them were separated by multiples of the comet's period or approximate half-period. In this connexion it may be recalled that Barnard found in the case of Holmes' comet a large irregular appendage covering an area of about a square degree and joined to the comet by a narrow hazy region. This comet, which moved in a fairly circular orbit ($e = 0.41$), showed extraordinary changes of size and brightness, and indeed was found during a sudden outburst of brilliancy in 1892. On the other hand, the idea has been put forward that the irregular brightening of Comet Schwassmann-Wachmann is related to sunspot activity, as the changes have at times coincided with the presence of a large spot or group of spots on that portion of the face of the sun turned in the general direction of the comet. But the plausibility of this suggestion is much reduced by the fact that the comet has also sometimes brightened without any exceptional sunspot activity being in progress.

There is accordingly good general agreement of the present theory with the known data on comets' tails. Long-period comets, for which e is very near to unity, produce prominent tails when nearing the sun, while short-period comets in orbits of moderate eccentricity show only very slight signs of tail-formation, and some none at all. Comets of intermediate periods with large eccentricity, like Halley's, continue to produce considerable tails at each perihelion return. A further qualitative feature in favour of this theory of tail-formation is the asymmetry of activity on the two sides of perihelion. Clearly the material first involved in collisions at and near the standard plane will tend to remain there and increase the probability of subsequent collisions. For this reason activity at any given point after perihelion would tend to be of greater intensity than at the corresponding point before it, a feature markedly associated with some comets.

Changes in dimensions of a comet

Since the great majority of comets have orbits of high eccentricity, the line of nodes NSN' for most particles will lie fairly near the major axis ASA'. As we have seen, this means that practically all the particles of such a comet will tend to pass through the standard plane somewhere near the perihelion point A. Those particles that fail to do so, but have their nodes nearer the ends of the latus rectum, LSL', will be so much fewer in number that they will be too sparsely spread to give any observable coma, and the observable extent of the comet when near perihelion will depend on the distribution of the particles concentrated towards the standard plane. Thus a good general idea of the change in size of a comet, measured perpendicularly to the orbital plane, can be obtained by supposing all the nodes to lie at A. In that case the width of the comet $2\delta H$ perpendicular to the plane at a general point $P(\theta)$ will be related to the half-width δH_B at the end of the minor axis B (see Fig. 12, p. 114) by the relation

$$\frac{\delta H}{\delta H_B} = \frac{PQ}{BC} = \frac{\sqrt{(1-e^2)}\sin\theta}{1+e\cos\theta}. \tag{4}$$

Since $\theta = 0$ corresponds to A, the formula gives theoretically a contraction right down to zero here, but this arises from the approximation involved in assuming all the nodes to lie precisely at A. In practice, as is shown by Table V, they will be distributed over a short arc round A, but nevertheless the formula will give an adequate approximation to the size of the comet (perpendicular to the plane) for other positions than those very near to A.

For long-period comets with e very near to unity, observations at B are not possible, but if instead the thickness $2\delta H_L$ at the end of the latus rectum ($\theta = 90°$) is taken as standard the corresponding formula reduces practically to

$$\frac{\delta H}{\delta H_L} = \tan \tfrac{1}{2}\theta. \tag{5}$$

This again gives theoretically a contraction right down to zero at perihelion, because of the simplifying assumption that all the nodes lie at A, so that in practice, as for (4), the present formula will not apply right down to $\theta = 0$.

Systematic measurements of the sizes of comas do not seem to figure prominently among cometary data, but a series of measurements are quoted for Encke's comet by Chambers.

126

In the following table are given first the theoretical values of $-\log \dfrac{\delta H}{\delta H_B}$ calculated from the above formula, and second the value of these expressions calculated from the observed radii.

TABLE VIII

Theoretical values of $-\log \dfrac{\delta H}{\delta H_B}$			Observed values of $-\log \dfrac{\delta H}{\delta H_B}$		
Position of comet, $\theta°$	Distance from sun (a.u.)	$-\log \dfrac{\delta H}{\delta H_B}$	Radius of comet (10^3 miles)	Distance from sun (a.u.)	$-\log \dfrac{\delta H}{\delta H_B}$
150	2·36	0·0	158·6	2·36	0·0
130	1·39	0·05	142·8	1·42	0·05
110	0·89	0·15	61·3	1·19	0·41
90	0·63	0·27	40·2	1·00	0·60
70	0·49	0·41	37·6	0·88	0·63
60	0·44	0·49	32·0	0·83	0·69
50	0·41	0·58	28·2	0·76	0·75
40	0·38	0·68	19·6	0·71	0·91
30	0·37	0·81	15·3	0·69	1·02
20	0·36	0·99	3·4	0·39	1·68
10	0·34	1·29	2·8	0·36	1·76
5	0·34	1·60	2·2	0·35	1·87
			1·5	0·34	2·02

The general agreement between these values is at least as satisfactory as could be expected in view of the number of likely sources of uncertainty. First, it is not stated by Chambers which radius was observed in each case, and as the comet may not have been even of approximately spherical form this renders the observational figures of no more than moderate value. Second, there might be expected some difference between the observed value and the relevant dimension perpendicular to the standard plane if this latter direction were not itself perpendicular to the line of sight from the Earth. Also, but by no means least in importance, the *observed* size, as already emphasized at a number of earlier points, would depend on the distribution of particles within the comet at each stage, and does not necessarily remain in constant proportion with the actual size. As a comet changes its shape, and at the same

127

time the space distribution of particles within it, the observable limit may not move with the particles. In view of these various effects, of which at present it is not possible to take any detailed account, it is plain that the comparison of the observed and calculated values of Table VIII can only be regarded as of a general nature.

There is evidently need for more explicit measures of the dimensions of comets, in so far as this is practicable, with a view to obtaining some idea of their dimensions referred to the orbital plane as a standard, and probably, where dimensions within this plane are concerned, referred to the particular direction of motion of the comet at the time of the observation and the direction perpendicular to this. Of course it would obviously not be possible to measure at any one time all three of these dimensions for any single comet, but for different comets, according to the positions of their orbital planes and the position of the Earth, it might well be possible to obtain some idea of how all three change as the comet pursues its orbit. There would of course yet remain the problem of how the observable size may be related to the actual size determined by the limit of the particle distribution, and the problem of how or whether the actual distribution can be inferred from the observations.

Expanding envelopes within a comet

We have referred earlier to the well-known observations of certain large comets which appear to show envelopes of material emitted from the central regions with almost explosive violence. These expanding shells, as they are sometimes termed, were particularly prominent in Donati's and Coggia's comets, and in Morehouse's comet of 1908. If these apparent motions transverse to the line of sight are accepted as real, it is found that repulsive forces from the centre of the comet of the most extraordinary strength have to be postulated, so strong indeed that early investigators were driven to suggest that perhaps comets were composed of matter of a kind unknown on Earth. Hypotheses of this kind are of course of no scientific value until it

128

PLATE XIII

A. Swift's Comet (1892 I) 1892 April 7, showing the tail emerging mainly from the central region of the head

B. Halley's Comet 1910 June 2 (30-minute exposure), showing the head with its surrounding fainter coma, and complicated tail structure

has been proved beyond doubt that all admissible hypotheses resting on known causes fail. In the present case, guided by the theory developed here, it seems far more likely that what is actually seen in these apparent expansive motions is the transmission of an effect through the comet. Collisions of particles crossing through the standard plane will not necessarily occur simultaneously at all points of the area of the standard plane occupied by the comet, and the 'expanding envelopes' might well be the advancing front in the standard plane of the colliding material coming from above and below the plane, probably as it becomes observable through being transformed into finer dust with its consequent far greater surface area available for reflecting sunlight.

It is also possible that the collisions would produce considerable heating effects especially at the surfaces of colliding particles. At a speed of 2×10^4 cm. sec.$^{-1}$ a particle possesses kinetic energy 2×10^8 ergs/g., but converted to thermal energy distributed *throughout* the material this would raise the temperature by only about 100°. But obviously the impacts would involve relatively very small areas of contact, and energy lost during the collisions might initially be communicated to only a small part of the mass actually involved. An action similar to that when flints are rubbed together to produce visible flashes must accompany the collisions, but with far higher speeds available in the case of the cometary particles. In such a way much greater heating than a mere 100° might result for a small proportion of the material, and it seems entirely possible that volatile constituents of the cometary particles would tend to be released through such collisional heating effects.

Changes in dimensions in the orbital plane

Still assuming that a comet consists of particles moving with identical periods, it is plain that as the orbital velocity changes the spacing of the particles parallel to the direction of motion will change in proportion. If we consider two particles X and Y describing the same orbit but at a distance δL apart, the time taken for X to move into a position formerly occupied by Y is

$\dfrac{\delta L}{v}$ where v is the velocity at the part of the orbit concerned.

If after a finite interval of time a pair of independently moving particles initially at X and Y have reached points X' and Y' a distance $\delta L'$ apart and have orbital velocity v', then we shall therefore have that

$$\frac{\delta L}{v} = \frac{\delta L'}{v'}.$$

If δL_p is the distance between the particles when in the immediate neighbourhood of perihelion, it is readily found that at a general point of the orbit

$$\delta L = \frac{\sqrt{(1 + 2e \cos \theta + e^2)}}{1 + e} \, \delta L_p. \tag{6}$$

For an actual comet, even though the period of every particle is the same, the individual orbit of any particular particle may have elements differing slightly from those of the standard orbit, except (in accordance with our present assumption) for the semi-major axis a. But clearly the above relation between the small quantities δL and δL_p, which can be regarded as measuring the extent of the comet along the orbit, will not be affected to the first order, and so will be applicable to the comet as a whole. It is important to emphasize once again, however, that the *actual* limit and the *observed* limit of any comet will not be expected to coincide, and that the relation between the two limits will not necessarily remain one of strict proportion. So the observational application of the above formula to any actual comet can again only be of a general nature.

TABLE IX. LENGTHWISE CONTRACTION OF A COMET ALONG ITS ORBIT

θ (degrees)	Encke	Halley	Parabola
180 (aphelion)	0·083	0·017	0·000
120	0·505	0·500	0·500
90	0·710	0·707	0·707
0	1·000	1·000	1·000

By means of this formula the values of $\delta L/\delta L_p$ at various points of the orbit have been computed for three different cases, first Encke's comet, second Halley's comet, and third the limiting values for a strictly parabolic comet (Table IX). For most comets observations would not be possible much beyond $\theta = 120°$ if as far, so that the actual observed change resulting from this cause is hardly likely to exceed a factor of 2. The change will take the form of an extension along the orbit as perihelion is approached. For orbits of small eccentricities, such as the annual comets 1925 II and Oterma, the ratio $\delta L/\delta L_p$ differs only very slightly from unity at most parts of its orbit, the maximum being about $1·3$ at aphelion for $e = 0·14$, for example. This particular change of dimension is obviously a purely periodic one involving no tendency for the leading and following particles of the comet to be lost, but it is to be remembered once again that the *observed* dimension would not be expected to change precisely in accordance with this relation.

Where the dimension in the orbital plane perpendicular to the direction of motion is concerned, the series of orbits of the various particles (still assumed to have identical periods) will be obtained by allowing small variations δe and $\delta \varpi$ to be made in the eccentricity and the angle measuring the direction of the major axis. If these adjustments are made, the resulting width of the comet, δN say, in the direction of the normal of the standard orbit, becomes an expression of so complicated a form that it has no obvious properties that could be interpreted as any simple change of dimension.

Observationally, the only comets for which comparisons of dimensions over a substantial portion of the orbit are possible are the short-period comets. Since, however, these have only small inclinations to the ecliptic the dimension δN in the orbital plane normal to the path would not usually be readily observable, whereas the thickness of the comet $2\delta H$ perpendicular to this plane would be the dimension most likely to present itself to observation at times more or less free from any projection effects.

Effect of slight differences of period for particles within the comet

Even supposing (as we have so far done) that at any one stage the particles of a comet all have precisely the same period, it is clear that collisions between some of its particles as they cross through the standard plane would usually reduce their total energy and thereby slightly reduce their periods. An estimate of the order of magnitude of the effect can be made from the values given on p. 119 for the angular size of a comet.

If two particles colliding at the standard plane have equal velocities before collision inclined at angle i to this plane, their velocity v will be reduced to $v \cos i$, assuming the collision to be inelastic. Since i is small the reduction may be taken as $\frac{1}{2} i^2 v$ with sufficient accuracy. A change in v of this kind at a fixed distance will change a because of the relation

$$v^2 = \mu \left(\frac{2}{r} - \frac{1}{a} \right), \qquad (\mu = GM)$$

which gives on differentiation

$$2v \, \delta v = \mu a^{-2} \, \delta a,$$

and hence that $\qquad \delta a = -v^2 a^2 i^2 / \mu.$

But since for most comets the collisions occur mainly at perihelion, we shall have

$$v^2 = \frac{\mu}{a} \cdot \frac{1+e}{1-e},$$

and hence $\qquad \dfrac{\delta P}{P} = \dfrac{3}{2} \cdot \dfrac{\delta a}{a} = -\dfrac{3}{2} \cdot \dfrac{1+e}{1-e} i^2. \qquad (7)$

Thus the period of such particles will necessarily be reduced, and after a further revolution of the comet they will accordingly be ahead of the main mass of the comet. They will become a whole revolution ahead after $-P/\delta P$ revolutions of the comet, that is after a time $-P^2/\delta P$. This quantity will therefore give an estimate of the time required for the present process to distribute particles right round the orbit. By (7), we have for this time

$$-P^2/\delta P = \frac{2}{3} \cdot \frac{1-e}{1+e} \cdot \frac{P}{i^2}.$$

132

For Encke's comet, $P = 3\cdot3$ years, $i \doteqdot 3 \times 10^{-3}$, $e = 0\cdot846$, and the time required is about 2×10^4 years.

For Halley's comet, $P \doteqdot 75$ years, $i \doteqdot 1\cdot5 \times 10^{-3}$, $e = 0\cdot9673$, and the time required is about 4×10^5 years.

These results give, of course, only rough order-of-magnitude estimates. In making them it has been assumed ideally that the collisions would be completely inelastic and head-on between equal particles, so that all the energy of motion transverse to the standard plane is lost. In fact, however, these conditions would rarely hold at a collision so this could seldom occur, and instead the collisions would result in a whole distribution of velocities about the foregoing assumed motion as a mean. Thus the actual values of δv would correspond to a more or less random distribution superposed on the previously assumed motion in which all the energy is lost. This would mean that most particles involved in collisions would undergo a decrease of energy and so the majority of them would have their periods shortened and accordingly thereafter advance ahead of the comet, but a small proportion would have their velocities increased and would move thereafter with slightly longer periods and so trail behind the comet. It will be understood that the changes in velocity here involved will be very small compared with the general orbital velocity, so the directions of motion of a particle subsequent to a collision will be practically in the original direction of motion of the comet, and the new orbit of the particle therefore will differ only infinitesimally from the general standard orbit, though now also in period.

Such particles, gradually moving out from the comet, can evidently be identified with the meteor streams known to be associated with many comets, which move in orbits similar to that of the comet concerned but usually show considerably greater lateral extent in the direction radial from the sun. The theory suggests that if this were the sole cause of meteor streams they should usually be stronger in the part *ahead* of the comet than in the following portion, but we shall see that there are other causes tending to spread the periods of the particles and thereby also bringing the meteor streams into existence.

133

Diversion of a comet from a long-period orbit to one of short period

We have seen in the discussion of the accretion process that small as cometary masses are they are nevertheless sufficient to exercise gravitational control over their constituent particles when at their initially great distances from the sun. This feature enables a limit to be set to the internal speeds that a particle can have relative to the comet as a whole without having enough energy to escape. For a comet of mass m_c and radius r, the energy per unit mass is of order Gm_c/r, and hence the velocities within the comet will be of order $\sqrt{(2Gm_c/r)}$. For $m_c = 10^{18}$ g. and with $r = 10^{11}$ cm. this comes out to about 1 cm. sec.$^{-1}$. This value is extremely small compared with the orbital speeds when the comet is moving entirely under the sun's influence, which are of order 10^6 cm. sec.$^{-1}$. During the time the sun's differential field predominates, a period say of the order of 100 years, the distance that particles might move out of the comet at a speed of 1 cm. sec.$^{-1}$ would be of order 3×10^9 cm., and this is fairly small compared with the size of a long-period comet (see Table VII, p. 119). Thus at the next aphelion passage such particles would still be within the gravitational control of the comet.

Let us consider next what effect is likely to be produced as far as the velocity-distribution among the particles is concerned when a comet is diverted into a short-period orbit as a result of a close encounter with Jupiter. The motion during the encounter can be regarded to a high approximation as a hyperbolic sweep around Jupiter of each of the particles of the comet. The final velocity relative to the planet of each particle when it recedes to a great distance from the planet will be precisely what it was before it entered the latter's sphere of action, but the directions of the velocities will be changed. Because of the finite size of the comet, however, each particle will not be turned through precisely the same angle, and second, although the velocity after encounter with Jupiter of each particle will be the same, the particles will not all be at precisely the same distance from the sun, and hence their individual total energies

134

(per unit mass) will not be precisely equal. This means that their periods will be distributed about a mean as a result of the encounter. The size of these effects and the resulting range of distribution can easily be estimated.

Let us consider first the angular deviation. If a particle starts from a great distance with velocity V and moves in a hyperbolic orbit round a centre of force μ/r^2 per unit mass, energy considerations give at once for the semi-major axis a

$$V^2 = \mu/a.$$

If the perpendicular distance from the centre of force (the planet) of the line of the initial velocity is p, then angular momentum considerations give

$$pV = \sqrt{[\mu a(e^2 - 1)]},$$

and it follows from these relations that

$$\sqrt{(e^2 - 1)} = pV^2/\mu. \tag{8}$$

Now the angle turned through, ψ say, during the motion is the supplement of the angle between the asymptotes, which is $2\tan^{-1} b/a$. Thus we have

$$\psi = \pi - 2\tan^{-1} b/a$$

$$= 2\cot^{-1}\sqrt{(e^2 - 1)}$$

$$= 2\tan^{-1} \mu/pV^2, \tag{9}$$

and this may be written

$$\tan \tfrac{1}{2}\psi = \frac{GM(\text{Jupiter})}{pV^2}. \tag{10}$$

Now we are concerned with the spread of direction, $\delta\psi$, consequent upon the variation δp of the distance from the planet of the initial velocity due to the finite size of the comet. This can be obtained by differentiating (10), thus

$$\tfrac{1}{2}\sec^2 \tfrac{1}{2}\psi . \delta\psi = -\frac{GM_J}{p^2 V^2} \delta p,$$

135

and by means of (10) this reduces to

$$\delta\psi = -\sin\psi . \frac{\delta p}{p}, \tag{11}$$

a relation that may be obtained directly from (10) by logarithmic differentiation. If we take $\delta p = 10^{10}$ cm. as the radius *before* the encounter and $p = 1/10$ a.u. $= 1{\cdot}5 \times 10^{12}$ cm. this gives even for large angles of deflexion ($\sin\psi \doteqdot 1$) that

$$\delta\psi < 7 \times 10^{-3} \text{ radian.} \tag{12}$$

For small angles of deflexion, which would also be associated with larger values of p, the value of $\delta\psi$ would be considerably reduced.

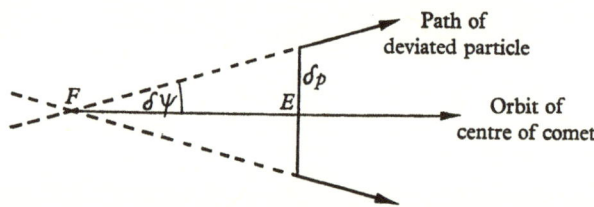

Fig. 14. Showing how particles in the neighbourhood of E deflected by different amounts will move approximately as if they emerged from the point F.

In the subsequent motion of the comet this spread of the directions of motion of its constituent particles will have an effect much the same as if all the particles were to have emerged from a point F at distance $\delta p / \delta\psi$ from the place E at which the encounter with Jupiter is effectively over (Fig. 14). Thus the motion of the particles will lie between directions inclined at angles $\pm \delta\psi$ to the intermediate direction followed by a particle moving at the centre of the comet. It is clear that in so far as the periods are unaffected (a point to be examined in a moment) the individual paths will diverge only temporarily, simply because all their subsequent elliptic paths round the sun must return to the neighbourhood of the point F. Thus the effect of such an encounter, deviating a long-period comet into a short-period orbit, would be to give the comet slightly increased size during part of the orbit, but not to an extent inconsistent with the observed dimensions of comets.

Turning now to the question of the effect on the periods, it is clear that after such an encounter with a planet even though the velocities of all the particles were precisely the same, their effective distances from the sun would now differ by amounts of order δp. As we have seen, for a comet moving at great distance in an almost parabolic path, each particle will be moving with practically the same velocity and as far as their orbits are concerned each particle can be considered as being at exactly the same distance from the sun. (The dimensions of the comet in the radial direction from the sun will not affect the question of the periods of the particles.) But after the encounter the effective radius δp of the comet will have been turned through a considerable angle, and particles now having transverse velocity components relative to the sun comparable with the radial components will find themselves at distances from the sun differing by amounts ranging from zero up to a length of order δp. From the energy equation

$$v^2 = \mu\left(\frac{2}{r} - \frac{1}{a}\right),$$

if v remains constant from one particle to another, we have that

$$2\frac{\delta r}{r^2} = \frac{\delta a}{a^2}.$$

In this r will be approximately Jupiter's distance from the sun and the new short-period orbit will have a value of a of the same order as r. For Encke's comet, for example, $a = 2 \cdot 2$ a.u. Thus we shall have in order of magnitude $\delta a \doteqdot \frac{1}{2}\delta r$, and hence where the variation of period is concerned

$$\frac{\delta P}{P} = \frac{3}{2} \cdot \frac{\delta a}{a} = \frac{3}{4} \cdot \frac{\delta p}{a},$$

where $\delta p(= \delta r)$ is the radius of the comet before the encounter. Putting $\delta p = 10^{10}$ cm., and $a = 5 \cdot 2$ a.u. $= 8 \times 10^{13}$ cm. this gives $\delta P/P = 10^{-3}$.

In the present case, if the particles near the centre of the comet are taken as standard, the variation δP of the period will be both positive and negative, so that particles will tend to be distributed round the orbit both ahead and following the central region of the comet. Hence the time taken for particles to be distributed right round the orbit will in the present case be of order $P^2/2\delta P$. For Encke's comet this comes to about 2×10^4 years, while for Halley's comet the time required is about 4×10^4 years. If these comets were, in fact, deflected into their present orbits by deviations through smaller angles these intervals would be somewhat increased, but on the other hand if the radius of the comet before the encounter were taken as 10^{11} cm. the times would be decreased by a factor of 10. It is obviously not possible to be definite about either of these factors, but there appears to be no observational evidence against the comparatively short former existence of Encke's comet in its present orbit. If we extrapolate backwards Holetschek's result, that its brightness is diminishing by 1^m per century, it would not be many centuries ago that Encke's comet would have been the brightest object in the heavens, and this suggests that the above estimates are probably not seriously in error.

The distribution of particles within a comet

For a long-period comet, when at great distance from the sun, it has been seen that the internal motions of its particles would be of order 1 or 2 cm. sec.$^{-1}$. Collisions will be infrequent and so every particle will be free to move under the attraction of the comet as a whole and describe some kind of orbit within the comet. The steady-state distribution of particles in such a system would probably be somewhat analogous to that of the stars in a globular cluster, increasing in number per unit volume towards the centre and having a symmetrical distribution. According to Eddington's studies of globular clusters it is possible that the distribution in such a system is representable by a law of the form $Ae^{-r'/r_0'}$ though with a cut-off at a reasonably large central distance, r.

138

In globular clusters the law $(1 + r^2/r_0^2) - \frac{5}{2}$ has been found to fit fairly well and this is very similar in shape to the curve of e^{-r^2/r_0^2}. The 'size' of the comet would depend on r_0 and be effectively a moderate multiple of it. It is also seen that the self-gravitation of the comet will not cause it to contract indefinitely, because of the small probability of collisions, but to maintain itself in a steady-state distribution in which each particle is in motion under the attraction of the comet as a whole. The time associated with such a motion will be of order $2\pi\sqrt{(r_c^3/Gm_c)}$, which comes to some thousands of years.

Now if a comet whose internal structure obeyed such a law were deflected by the action of a planet we have seen that some particles would be speeded up and others slowed down relative to the average, and it is reasonable to expect that the new distribution will still be closely representable by a similar law, but with a value of r_0 now slightly increased. If such a distribution is even approximately correct it enables us to form an idea of how a comet moving in an orbit of moderate period but high eccentricity, would develop as it approached the sun. When first visible it would probably be seen at a place beyond the end of the latus rectum of its orbit, and so the radius of the distribution would be nearly at its greatest value. The number of particles plotted against the apparent distance from the centre would be distributed as in Fig. 15 (i). The comet would be visible out to a certain distance CR at which the number of particles in the line of sight, which is proportional to RS, is still sufficiently great. Now as the comet approaches more closely to the sun, the greater proportion of the comet contracts, because of the concentration of the nodes near perihelion, and the distribution becomes correspondingly more closely concentrated round C the centre, as in (ii). The corresponding radius would now appear to be CR', where $R'S' = RS$. Thus the general contraction of the whole distribution, involving a decrease in the parameter r_0, would result in a contraction of apparent radius, but there is no reason to expect that the two would remain in strict proportion. The actual contraction would depend on the eccentricity of the orbit, as we have seen, since it is this that largely settles the degree of concentration of the nodes near perihelion, but the apparent radius will also depend on the precise position of the Earth since at different distances the necessary effective depth of particles for observation of the comet to be possible will be different.

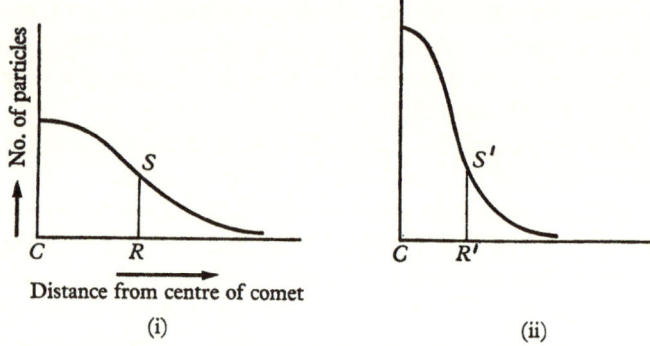

Fig. 15. Distribution of particles of a comet at different distances from the sun. The number of particles in the line of sight is proportional to the ordinate RS.

In (i) the comet is at greater distance from the sun than in (ii), and the requisite limiting depth of particles may occur at greater distance CR from the centre of the comet than the corresponding value CR' of diagram (ii).

Finally, there is the possibility that the actual distribution may not by any means be spherically symmetrical, for we have seen that slight differences of period will tend to spread the particles along the orbit, and this would mean that in these directions the parameter of the distribution (measured by the distance over which the number of particles per unit volume falls to a definite fraction, say half, of the central value) may be much larger than in directions at right angles to it. For such a distribution, the apparent shape as seen from the Earth, depending as it would on the number of particles per unit cross-sectional area in the line of sight at any part, would only bear a general relation to the distribution of particles. The very fact that the distribution in the direction of the orbit may be widely spaced would tend to make the volume-density of particles appear smaller in these directions and the comet might be observed as far smaller than it actually is. The truth of this can be seen at once by imagining the particles to be distributed over an increasingly great area: at a certain stage the particles will be so far apart that the comet would become invisible, yet the actual extent would be very large. We are accordingly precluded from knowing for certain what the actual dimensions of comets may be. The only immediate hope of ascertaining this appears to be that the Earth might sooner or later chance to pass so near the centre of a comet that direct observations on the resulting meteor showers could be made.

The dynamical decay of comets

It is clear that the effect of slight variations in period causing particles to spread around the orbit must be to diminish gradually the volume density of particles near the centre of a comet, and it must be supposed that through this cause comets will become steadily diffused into meteor streams around their orbits. There is abundant observational evidence supporting this. In the first place most short-period comets are faint, and presumably therefore of small mass compared with long-period comets. When one of the latter is first diverted into a short-period path, probably only a small fraction of its particles are deflected into orbits of sufficiently nearly identical periods for them to continue to constitute an observable aggregate. The remainder would soon spread out in the general direction of the orbit, some going on ahead and some following behind. Also the process of spreading out, as we have seen, proceeds more rapidly the shorter the period of the comet, and this again is consistent with observations of meteor streams. Indeed, there are some short-period streams recently discovered by Lovell and his colleagues for which no associated comet is known.

There is, however, another effect which must tend gradually to dissipate a comet, namely the so-called Poynting-Robertson effect due to the braking action of the sun's radiation. The nature of its mechanism is easily understood in general terms.

141

If we imagine a small particle moving in an orbit round the sun, it is constantly absorbing light received directly from the sun and re-emitting it in all directions symmetrically round itself. Now the radiation flowing from the sun carries both mass and momentum, but practically no angular momentum because the sun has negligible rotation. On the other hand, when this light is re-emitted, the particle effects this in its own proper frame, and so the light then has angular momentum per unit mass about the sun the same as that of the particle at the time of re-emission. This angular momentum is necessarily supplied by the particle, which itself must thus gradually lose angular momentum, and so the parameter of its orbit, $a(1 - e^2)$, must slowly decrease. For an object like a planet the mass of the emitted radiation is so small that the effect is utterly negligible, but for a particle of sufficiently small radius for which the area is large in comparison with its mass the effect rises to great importance. Thus Robertson has shown that a particle of density σ g. cm.$^{-3}$ and radius a cm., moving initially in a circular orbit at distance R a.u. from the sun, would be swept into the sun in a time

$$7 \times 10^6 \sigma a R^2 \text{ years.}$$

Thus if $a = 10^{-4}$ cm. and $\sigma = 2 \cdot 5$ g. cm.$^{-3}$, the time taken from the Earth's distance would be less than 2000 years, and for $a = 10^{-3}$ cm. less than 20,000 years.

Since a comet, according to the present theory, is wholly constituted of small dust particles, it is clear that we have in the present process a further effect tending to reduce the period of motion of its particles, and hence tending to accelerate the smaller ones ahead of the comet along the general direction of the orbit. Thus in this braking effect of solar radiation we have yet another cause tending to produce meteor streams.

The acceleration of Encke's comet

The effect of solar radiation in retarding the motion of small particles has been advanced by a number of writers as a possible explanation of the observed acceleration of Encke's comet

(Chapter I, p. 4). The strength of the effect is different for different-sized particles, smaller ones being much more retarded than large ones, and therefore the smallest particles would tend to be accelerated ahead (the retardation produces an angular acceleration because of the loss of angular momentum) more rapidly than the larger particles. Since what is observed for any comet is presumably the central more concentrated region of a much larger distribution of particles, this distribution will itself be shifted in the forward direction, but it is by no means clear how the motion of the observed centre will be related to the effect. Nevertheless, it seems possible that an acceleration will result, and this has been tacitly assumed by all earlier investigators.

On the simple assumption that the comet as a whole undergoes an exactly analogous acceleration to that produced in a single particle, Robertson (1937) has shown that the first-order perturbations of the various orbital elements are related to the loss of angular momentum $-\delta h$ per revolution in the following way:

$$\frac{\delta l}{l} = -2\frac{\delta h}{h}, \quad \frac{\delta e}{e} = -\frac{5}{2}\frac{\delta h}{h}, \quad \frac{\delta n}{n} = \frac{3}{2}\cdot\frac{2+3e^2}{1-e^2}\cdot\frac{\delta h}{h} = -\frac{3}{2}\cdot\frac{\delta a}{a}. \quad (13)$$

Since for Encke's comet $e = 0.8458$, this gives for the proportions of the several variations

$$\frac{\delta l/l}{-2} = \frac{\delta e/e}{-2.5} = \frac{\delta a/a}{-14.6} = \frac{\delta n/n}{21.9}, \quad (14)$$

but according to Robertson these values cannot be reconciled amongst themselves with the observed changes; in particular the variation in the parameter, δl, is not consistent with the value inferred from the above equations with the observed value of δn, the change in mean motion. This conclusion had also been reached much earlier by Plummer (1906) but with incorrect values for the effect of radiation pressure.

The present theory of the formation of cometary tails by collisions near perihelion introduces an additional effect that will tend to accelerate the comet. If we consider two equal

143

particles symmetrically placed on different sides of the standard orbital plane, so that their orbits before collision are inclined at angles i and $-i$ to this plane, then their angular momentum vectors before collision will be inclined at these same angles to the normal to the standard plane. If the collisions occur at perihelion, as we have seen may be assumed, changes in the orbital elements of these particles will be such that $a(1-e)$ remains unaltered, while the velocity there and angular momentum are reduced by a factor $\cos i$. According to the theory, only a small proportion of the mass of the comet will be involved in collisions at each revolution, but the general effect of the process on the motion of the centre of the distribution of the particles may be expected to be that of a decrease of h without any change in $a(1-e)$. On this basis, from the usual formulae for elliptic motion it is readily found that the corresponding relations to (13) are now:

$$\frac{\delta l}{l} = -2\frac{\delta h}{h}, \quad \frac{\delta e}{e} = -\frac{2}{e}(1+e)\frac{\delta h}{h}, \quad \frac{\delta n}{n} = \frac{3(1+e)}{1-e} \cdot \frac{\delta h}{h} = -\frac{3}{2} \cdot \frac{\delta a}{a},$$

(15)

and for $e = 0.8458$, the numerical values of the ratios of the variations are now

$$\frac{\delta l/l}{-2} = \frac{\delta e/e}{-4.36} = \frac{\delta a/a}{-23.9} = \frac{\delta n/n}{35.9}.$$

(16)

It will be seen that these are substantially different from the values given by (14) resulting from radiation pressure; and for a given acceleration, $\delta n/n$, the corresponding change in the parameter of the orbit is much smaller than in the former case. There is no doubt that the present process goes some way towards resolving the difficulty, but it is doubtful if detailed agreement can be attained, or indeed should even be expected at this stage, because of the difficulty in deciding what the actual observations mean in relation to a changing distribution of particles within the comet. Of course, internal collisions cannot alter the angular momentum of the comet as a whole, but they will tend to cause a flattening of the comet towards the standard plane by destroying the component velocities perpendicular to this plane. Both the radiation effect and the collision effect will operate more strongly for small particles than for large ones, but until more is known not only of the space distribution of the particles but also of the distribution of sizes, there seems no way that the problem presented by the acceleration of Encke's comet can be taken much further.

The action of the Poynting-Robertson effect in causing small particles to lose angular momentum at an appreciable rate and gradually spiral towards the sun represents the final stage of the complicated process by which interstellar dust captured by the sun, but not immediately falling into it, is nevertheless eventually swept into that body. Beginning as small solid condensations in gaseous material ejected from stars, the interstellar dust particles first form into diffuse but far denser aggregations as comets, and of those that happen to be deflected into short-period orbits the gradual dispersal into meteor streams is accomplished by internal collisions and any other disturbing effects as well as the drag due to solar radiation, while this latter mechanism also continually operates to draw the particles into the sun.

The development of sun-grazing comets when near the sun

It has been assumed all along, in conformity with the proposed mode of origin, that a comet consists of a large number of individual particles preserving their separate identities, apart from possible disruption into smaller fragments during collisions. But if a comet passes within a few radii of the sun's surface, as some comets are observed to do, and as practically all comets must do initially according to the present theory, the equilibrium temperature due to the sun's radiation becomes comparable with, though of course always less than, the surface temperature of the sun itself. This raises the question whether the particles of a sun-grazing comet can in fact remain in solid form during the sweep round perihelion. The answer would seem to be that they definitely cannot do so, for Russell has shown that all small particles would be completely vaporized if they passed within about four solar radii of the sun's surface, at which distance they would take up an equilibrium temperature of about 2000°. Particles moving in almost parabolic orbits and crossing within such a distance would spend a period of a few hours in so doing, and this would be ample time for the whole of their material to be raised to an equilibrium temperature. Indeed, it seems fairly certain that even very large particles, say of the order of several metres in radius, would be entirely vaporized during such a time if subjected to the sun's radiation, because vaporization at the surface would proceed so fast that the underlying material would soon become exposed and then be vaporized in its turn. There would be no dependence on conduction for the heat to be communicated through the body. But this point is scarcely of importance, however, for there is no reason to suppose that a long-period comet in its first sweep through perihelion would contain anything but extremely small particles.

Accordingly there seems no doubt that any sun-grazing comet must be completely vaporized during the few hours it spends near the solar surface. The extension along its orbit and sub-division into several parts observed for the great comet of 1882 must have taken place while the comet was entirely in gaseous form. This would not be in any way inconsistent with the fact that the comet was quite invisible when it transited the disc of the sun. For its light-stopping power would be no greater in the vaporized form when its density would be at most 10^{-12} g. cm.$^{-3}$. A square centimetre section through the centre of the comet would contain less than 0·1 g. and in gaseous form this would produce negligible obscuration.

The thermal velocities of atoms and molecules in a gas at such a temperature are readily estimated, for in a gas at temperature T a particle of molecular weight μm_H, where m_H is the mass of the proton, has thermal velocity

$$\sqrt{(3kT/\mu m_H)},$$

where k is the molecular gas constant ($1·372 \times 10^{-16}$ erg/degree).

For $T = 2000°$, and $\mu = 50$, this leads to a value of about 1 km. sec.$^{-1}$. But the orbital velocity of a sun-grazing comet near perihelion would be about 500 km. sec.$^{-1}$, so that any tendency of the gaseous material to expand because of its high temperature would lead to only a small divergence away from the standard orbit measured by an angle of the order of 2×10^{-3} radian. This is of just about the same order as the angular deviations already found associated with other sources of disturbance of a comet. At any rate, it would obviously be insufficient to cause any serious dispersion of the comet during the few hours that its temperature would be maintained at a value of order 2000°.

As soon as the comet recedes again to more than a few solar radii from the sun, however, the temperature of its gaseous material will fall, and the material will condense again into small particles, just as raindrops form in the atmosphere, simply because the pressure within such material when gaseous (but cooled) would far exceed the vapour pressure of material in the solid form at the corresponding temperature. The importance of this process in cosmogony was first pointed out by A. L. Parson in an important paper in 1946. It has been further studied by Hoyle, who has shown that the time taken for a particle of radius r cm. and specific gravity σ g. cm.$^{-3}$ to develop within a supersaturated gas of density ρ g. cm.$^{-3}$ at temperature T is given by the approximate formula

$$\frac{\sigma r}{\rho} \bigg/ \left(\sqrt{\frac{\mu m_H}{3kT}} \right). \tag{17}$$

For $T = 1000°$, $\mu = 50$, $\sigma = 2\cdot5$ g. cm.$^{-3}$ and $\rho = 10^{-12}$ g. cm.$^{-3}$ this leads to a value of

$$3\cdot6 \times 10^7 r \text{ sec.}$$

Thus particles of radius 10^{-4} cm. would grow in $3\cdot6 \times 10^3$ sec., that is one hour, while particles of radius 10^{-3} cm. would require about ten hours. Accordingly the recondensation of the material into solid form would proceed very rapidly once the comet had receded beyond a distance of a few solar radii.

147

Moreover, self-gravitation of the comet as a whole may come into play again when the comet approaches aphelion.

Of course it is entirely possible that a small proportion of the material, especially atoms and molecules of low molecular weight, would escape in gaseous form and leave the comet at such low density that solidification would not occur, but the continued existence of the comet itself would require only a proportion of the material to re-condense. Also we have seen that the thermal velocities within the comet would be small and therefore the thermal expansion of the comet as a whole during the few hours of its existence near perihelion would bring about a negligible decrease in density, and accordingly suitable conditions for the return to the solid state would be maintained.

There emerges from all this the strong inference that comets must be composed entirely of small particles. For not only do they form initially in the accretion stream from small dust particles, but complete vaporization at the first perihelion passage must also occur for all comets, with the exception perhaps of a very few that may happen through exceptional perturbation to attain sufficiently large perihelion distance during their first approach to the sun. This conclusion is certainly borne out by evidence from meteor showers. The proportion of very bright meteors is negligible, and there is no evidence whatever of which we are aware of any correlation between such showers and the fall of meteorites. If comets contained even a small proportion of their mass in the form of large masses, comparable in size with small meteorites say, strong showers would be expected to contain at least an occasional meteorite. Evidence from the structure of meteorites themselves also confirms this view, for such evidence shows that they must have been formed at very high pressures, of the order of those to be found only in bodies as large as planets, whereas the pressure within a comet even when at a temperature comparable with that of the solar surface would not be as much as one dyne per square centimetre.

Instability of a gaseous comet

The dynamical condition for instability of a gaseous or liquid mass moving under the sun's attraction can be applied directly to a comet in gaseous form. The comet will be unstable if

$$M/m_c \gg (R/r_c)^3,$$

where M is the mass of the sun, m_c is the mass of the comet, R its distance from the sun, and r_c the radius of the comet. Now even if m_c were as large as 2×10^{21} g. the left-hand side would be of order 10^{12}, whereas on the right-hand side R/r_c could hardly much exceed 10 for a sun-grazing comet. Even if R/r_c were 10^2 or even 10^3—impossibly high values—the condition would be strongly satisfied, and it follows that a gaseous comet would be highly unstable in that there would be no even approximately fixed boundary figure that could contain it. Some rapid relative dynamical motion must accordingly occur as the comet moves through the strong solar tidal field, and there can be little doubt that the remarkable elongation of figure and subsequent disruption of the great comet of 1882 resulted from this cause. The theoretical investigation of the details of the motion by which such an unstable gaseous system would develop presents great mathematical difficulties, but it was conjectured by Jeans through his researches on the problem that a filament containing a number of condensations might form in such circumstances. The problem that he had in mind was the formation of the planets; nevertheless, possibly an actual instance of these conjectured motions was provided by the 'string-of-pearls' development of the 1882 comet.

THE RELATION OF THE PRESENT THEORY TO EARLIER IDEAS

In presenting the theory of the foregoing chapters I am aware that there is room for further investigation at many if not all points. But it is in the nature of science, and indeed the great merit of it, that when a substantially correct or valid hypothesis is obtained, instead of closing up the subject and completing our knowledge of it, the very opposite happens and we are confronted with numerous unanswered, though not necessarily unanswerable, questions that beforehand could not even be raised except perhaps in the vaguest terms. In attempting to assess at this stage the present theory of comets a number of considerations have to be borne in mind.

First to be remembered is the inherent difficulty of all cosmogonical problems, both from the mathematical point of view and from that of deriving valid hypotheses. The relevant processes must essentially involve questions of redistribution of the matter of the universe. A system beginning in a stable condition gradually evolves, or through external influence is caused to change, to some other state, and a passage from one stable form to some other form will result. Such a course of development must involve instability and dynamical processes of an irreversible nature, and nearly always will involve motion in the neighbourhood of neutral equilibrium conditions. As is well known, the mathematical difficulties associated with such questions are usually so great that apart from tracing the development up to the point where instability sets in, only considerations of a general nature can be advanced in attempting to decide the future course of the motion. The comet problem proves to be no exception to this. The motion of the cloud near the sun is, as we have seen, of a highly unstable character and subject to large possible changes in its details, but not in its general form, through any slight external disturbances such as must always be assumed to be present. Then again, within the accretion stream gravitational instability is an essential feature of the later development of comets. Thus at both its important stages we have in the process, what is indeed an essential requirement for the phenomena to be explained, a highly flexible mechanism subject to a large possible range of variation in the products that emerge from it, while all the time the mechanism itself preserving its same general character. But clearly it would not be consistent with observed requirements if the theory led to a perfectly rigid development of regular dense masses, for this is not what comets are like.

Second is to be recalled the position that has existed for so long in relation to comets where the theoretical cosmogony of the solar system is concerned. Indeed, even in Jeans' works no mention of any kind of comets is to be found despite his great imaginative resource in speculative fields. Now any scientific theory is properly to be regarded as a developing structure, rather than as an instantaneously created finished product that immediately makes clear the answers to all questions and renders further discussion unnecessary. There may exist branches of knowledge in which this latter possibility holds, but if so they must indeed be remote from science. Therefore it would not be reasonable to expect that the development of the present hypotheses on the origin and subsequent evolution of comets, largely correct as the discussion in this book shows them to be, is now complete or indeed can ever approach completeness. What we do have, however, is a definite set of ideas concerning the origin and structure of comets that links them up to other astronomical systems, and to which can be addressed, and indeed themselves suggest, a large number of new questions concerning comets, and moreover bid fair to making it possible to answer such questions in terms of known processes. We can perhaps best make clear this aspect of the theory by comparing the position with that provided by earlier 'theories' concerning comets.

Where the origin of comets themselves is concerned, if we exclude Aristotle's suggestion of a terrestrial origin and location for them, ideas appear to have been of three different types. But none of them can properly be described as rising to the level of what a scientific hypothesis should be (though in fact they nevertheless have been so described). In the first place it has been suggested that comets are expelled from the bodies of the planets. This idea, such as it is, seems to have been associated principally with the name of R. A. Proctor, though its originality and ingenuity are so evanescent that it can hardly have escaped occurring to anyone else until 1870, the time of its popularity. No doubt numerous astronomers before Proctor conceived the idea, but it would require an exceptionally uncritical attitude to overlook its patent objections and weaknesses. First we see that it is not a hypothesis in the scientific sense, for it has no foundation either in observation or theory, quite apart from the subsequent question whether it fits in with known properties of comets. If a single comet were ever to have been observed emerging from a planet this would provide a valid basis for the idea, or if the theory of the structure of planets were to show that expulsion of part of their mass was possible in a form capable of being identified with a comet, this again would afford a reasonable basis for the idea, and it would become at once a legitimate hypothesis. But no such evidence exists, nor was any attempt made to provide the necessary basis. From the point of view of scientific method any such idea should in the first instance be developed by considering the theory of the structure of planets before making it the basis of any further speculations.

As soon as we attempt to discuss the idea in terms of what is known about comets and planets, it is found to involve manifold difficulties. First, if comets had emerged in some way from planets, the association of their orbits with the planetary orbits would be far stronger than it is, since apart from extraneous disturbances each cometary orbit would be extremely closely related to that of its parent planet. The few comets that are so associated represent a negligible proportion of the whole vast

152

number, but even for these the closest distances from Jupiter's orbit is very great compared with what it should be, at least for some, if the comets had actually been ejected from the planet. Then again, nearly parabolic orbits would be the exception and not the rule for such an origin. If on the other hand we turn to the planets themselves, only Jupiter can by any extension of thought be reckoned as the source. But for small solid particles to be driven upwards through the vast atmosphere of this planet would require absurdly high initial speeds, and we know that even in the Earth's much shallower atmosphere meteors (themselves parts of comets) are completely vaporized. This alone makes it clear that there is no possibility of anything remotely resembling a comet coming out of a planet. The requisite high speed would cause it to vaporize at once, and a jet of gas at high temperature fired out of Jupiter, even supposing it could penetrate the atmosphere, which is highly doubtful, would probably dissipate as soon as it got into the empty space beyond. Then again, many comets have far greater overall bulk even than Jupiter, and a few greater than the sun itself, whereas one would expect that anything shot out from a planet would if anything be considerably smaller in size than the source. Finally, there is nothing in the whole theoretical knowledge of the structure of planets to suggest the ejection from them of masses with hyperbolic speeds. On all these several counts the idea utterly fails, and is seen to be nothing more than an ill-conceived conjecture that could only be kept in countenance by ignoring all known properties of both planets and comets. Unfortunately for the progress of science ideas of this sort are commonplace in astronomy and their authors, particularly by their insistence in terming them 'theories', bring untold discredit to the subject. Genuine theoretical work is often lumped in together with barren suggestions of the crudest inapplicability with the result that the latter are available to be instanced as characteristic of all theoretical work.

Possibly guided by his hypothesis of a planetary origin for comets, Proctor remained strongly opposed to the 'capture' hypothesis for the short-period comets, according to which they have been deflected into their present orbits by Jupiter. It was argued by Proctor that the gravitational sphere of action of a planet such as Jupiter is so small compared with that of the sun that the probabilities of capture of comets into the present orbits are far too small to explain the number of existing period comets—a number of order 100 out of a total of perhaps 200,000 comets. But Proctor's argument fails to take account of the fact that it is not necessary for a long-period comet to be deflected at once, at its very first encounter with a planet, into a short-period orbit, and that what is far more likely is that it will first be deflected into an orbit of moderate period of a few hundred years, an orbit which of necessity will pass near that of the planet concerned (Jupiter) and so lead eventually to further deflexion. The probability of these additional encounters is far greater for an already captured comet than for one moving at random, and as was shown by H. A. Newton they will operate to bring about very considerable increase in the number of comets with short period, direct motion, and small inclinations. According to Newton's calculations, out of 100,000 comets that enter a sphere centred at the sun of radius 1 a.u., the numbers that are changed by a single encounter into comets visible from the Earth and of period less than a thousand years by the action of Jupiter will be about 90, by that of Saturn less than 3, and by the remaining great planets a negligible proportion. These figures as they stand already suggest a probability approaching the right order of magnitude, and since practically all comets are in a sense periodic, there will be further opportunities for deflexions into orbits of shorter period. The process must be regarded as going on continuously and gradually producing new short-period comets. The existence of some such process within the solar system is of course a necessity since short period comets can scarcely have lifetimes, as comets, of more than a few thousand years.

The fact that a number of instances of large deflexions of known periodic comets by the planet Jupiter have taken place during the past century is itself indicative of the fairly high probability of such changes, but there is the possibility of further cumulative change occurring through planetary action without any actual close encounter taking place. The theoretical investigation of such changes in any particular case would be an extensive problem of computational celestial mechanics which has not yet been attempted. Observational evidence of such changes, however, is provided by Comet Pons-Winnecke whose osculating orbital elements even over the past hundred years have shown considerable secular changes, as the following table shows:

TABLE X. OSCULATING ELEMENTS OF COMET PONS-WINNECKE

	1858 II	1886 VI	1898 II	1927 VIII	1933 II	1939 V	1945 a
q (a.u.)	0·764	0·886	0·924	1·040	1·100	1·100	1·160
e	0·755	0·726	0·715	0·685	0·672	0·669	0·655
P (yrs.)	5·560	5·800	5·880	5·990	6·160	6·080	6·150
ω°	162·000	172·000	173·000	170·000	169·000	169·300	170·100
Ω°	113·000	104·000	101·000	98·000	97·000	96·800	94·500
i°	10·800	14·500	17·000	18·900	20·100	20·100	21·700

It is clearly possible that even more rapid changes may occur from one revolution to the next in orbits of comets of moderately long period, but obviously a sufficient amount of observational evidence has not yet accumulated for the direct investigation of this question.

On a different level altogether from Proctor's ideas on the origin of comets was the suggestion by Chamberlin (1901) that comets might be produced by the disruption of small asteroids passing within the so-called Roche limit (p. 49) of the great planets. The existence of over 1500 asteroids already discovered suggests that very many more are present in the solar system either at greater distances than any of those yet observed or having far smaller mass and size than them, in addition to those comparable in size and position with the known ones, but not yet discovered. Baade has estimated that the number that may be capable of being discovered by the 100-in. telescope at Mount Wilson exceeds 30,000, but as there may in fact be no definite lower limit of size the actual number may be far larger than this. At all events, as far as the existence of numerous asteroids is concerned, Chamberlin's hypothesis probably rests on fairly safe ground. A serious difficulty is met with, however, in explaining comets of high inclination moving in orbits that do not pass near that of any planet, and indeed few comets at present have orbits that pass within anything approaching the Roche limit of Jupiter. The hypothesis, if it were admissible, might explain some comets, but not all. Where the mechanism itself is concerned, however, some disruption of an asteroid by gravitational action appears to be a dynamical possibility but cohesive forces would place a serious lower limit to the general size of the resulting fragments. According to calculations by Jeffreys, an asteroid 200 km. or less in radius could graze the surface of Jupiter without undergoing disruption owing to its internal cohesion, and even if this figure were only correct in order of magnitude its derivation makes clear that gravitational disruption into pieces as small as meteors is an utter impossibility.

An idea somewhat similar to that of Proctor is contained in the suggestion due to Crommelin that comets have a solar origin and are in some way associated with prominences. Few astronomers have studied comets from the observational standpoint more assiduously than did Crommelin, and if ever the futility of the direct observational approach to theoretical problems were clearly demonstrated the present suggestion remains as a monument. It seems scarcely worthwhile even drawing attention to the absurdity of the conjecture, but unhappily the literature of astronomy shows that almost equally absurd notions are still actively entertained in regard to many other problems, besides that of the origin of comets. At the time that Crommelin advanced the idea, however, the nature of prominences could scarcely even be guessed at, so that his idea merely displaced the origin of comets to a realm in which no reasonable discussion of the conjecture could be conducted at all. This in itself is opposed to the correct scientific attitude in such questions. It should perhaps be mentioned, in fairness to Crommelin, that he himself considered this 'explanation' of the origin of comets to be inadequate for those with perihelion distances greater than the Earth's orbital radius, which means rather more than 200 solar radii, though even so it is difficult to understand this curious piece of caution, for if perihelion distances out to, say, 100 solar radii are explicable by the process (though it is not here suggested for a moment that they are) then there seems little or nothing against perihelion distances of 1000 solar radii, or indeed, any value.

Yet another suggestion for an origin within the solar system is that comets represent the debris remaining after the planets had been formed, that is, are portions of a hypothetical original primitive solar nebula. But such an idea is opposed to some of the simplest and most dependable features of cometary orbits. The uniform and common direction of motion of the planets and minor planets round the sun indicates that the same feature must have characterized the distribution of the material at an earlier stage. But the arrangement of the cometary orbits is quite inconsistent with this, as also are their strongly parabolic forms. Laplace himself, it should be said, was firmly convinced that the planets and comets demanded different origins. Further, we would expect all bodies formed with the planets to be of planetary character, but if we overlook this and admit otherwise, it would still need special additional explanation why two types of object should have formed. But then, from a different standpoint, the rates of disruption of comets appear to be far too rapid for them to have existed anything like as long as the planets, and it is doubtful if an age as great as 10^8 years can be supposed even for long-period comets.

More absurd still than the planetary origin is the suggestion that comets represent an earlier stage of development of asteroids. As mentioned earlier, there are some 1500 known asteroids, all moving in orbits of moderate inclination to the ecliptic and every one, without exception, pursuing direct motion.

157

This alone makes the present idea quite an impossible one, for it would need additional suppositions to get from the randomly distributed cometary orbits to the regular arrangements of the asteroid orbits in order to give it a vestige of tenability. From the physical standpoint, there is no resemblance at all between asteroids and comets apart possibly from their small masses— though Ceres, for one, is at least a thousand times, and probably a million times, more massive than even a large comet. If any connexion exists there should exist meteor streams associated with asteroids that come within the Earth's orbit—Adonis, Apollo, and Hermes—or again further explanation is needed why none should be found. Certain it is that the authors of the suggested idea have pointed out no evidence of any kind supporting it, nor have they emphasized the cogent evidence adverse to it.

Having in turn exhausted every source within the solar system, a series of no less absurd speculations have been advanced for an extraneous origin. Lest it be thought that in writing of these suggestions an undue attempt is being made to belittle them and consign them to a plane below their true worth, we quote here two *alternative* answers to the riddle of the comets in words made use of by their originator:

'*Either* (i) comets are chance visitors wandering through space and now and again casually caught up by the sun, or by some of the major planets acting like the old naval press-gang and compelling them to attach themselves to the sun and by taking elliptic orbits to become permanent members of the solar system.

Or (ii) they are aggregations of primeval matter not formed by the Creator into substantial planets, but left lying about in space to be picked up and gathered into entities as circumstances permit.'

Perhaps a more remarkable phenomenon than even the comets themselves is the fact that such notions can be seriously advanced as constituting a contribution to the resolution of the problem.

158

The general hypothesis that comets are simply overtaken by the sun in its galactic motion is disposed of at once by the fact that no truly hyperbolic path is known, whereas all paths would be initially hyperbolic if this were their origin, as Carrington correctly pointed out as long ago as 1860. This difficulty has not prevented numerous attempts to revive the hypothesis. The mechanism of three-body capture, through the combined influence of the sun and Jupiter, might have something to be said for it, despite its small prior probability, were all comets closely associated with Jupiter, but by far the majority of comets have orbits bearing no relation to that of the planet at all, and the hypothesis could therefore at best account for but a negligible few, even if we disregard the fact that those comets that are associated with Jupiter can more readily be explained as having been deflected into their present orbits from nearly parabolic long-period orbits already gravitationally attached to the sun.

Once again, the notion that comets are simply picked up from interstellar space is no more than a displacement of the domain of the problem to a vaguer realm, for it in no way answers the question how such curious objects, with their very low mean densities by planetary standards but very high densities by interstellar standards, could have come to form. There is no observational evidence of such concentrations within interstellar dust clouds, indeed, there is definite theoretical evidence against the idea, for the shearing motion in the galaxy would effectively operate to prevent any such aggregations forming. On the other hand, H. N. Russell, in discussing the general question of the origin of the solar system, has put forward a more plausible though still very imperfect modification of the capture idea. His account of the mechanism is only in descriptive terms, but it bears some degree of resemblance to certain broader features of the accretion process as discussed in the foregoing chapters. Thus Russell discusses the possibility of the sun passing through a nebula, and actually writes of 'a decided temporary concentration in the axial line behind the sun', but he goes on to suppose that a gaseous cloud would

159

result, associated with the sun but almost stationary behind it, and that this cloud would then offer resistance to the motion of comets initially describing hyperbolic orbits round the sun, and thereby possibly converting some of them to elliptic ones. The comets, Russell considered, would have to have been already present in the nebula as condensations, and moreover in sufficient profusion for the process to explain the huge number left behind in the solar system. His discussion of the idea was very brief and not such as to suggest that he had any great confidence in it. There is now no difficulty in seeing numerous fatal objections to the theory even in this vague undeveloped form, and certainly Russell himself advanced it only with due apology in default of any other reasonable mechanism. The account fails to perceive that the convergence of material towards the axis will itself bring about just the requisite degree of concentration of the particles, and moreover simultaneously effect the conversion of hyperbolic to elliptic motions.

Yet another obscurant idea dating back indefinitely is that a vast assembly of comets associated with the sun subsists in the form of a nebulous shell at 10^4 or 10^5 astronomical units distance (where it would be quite beyond the possibility of observational detection), and that the few comets seen are rare members of this cloud that happen to be deflected inwards to a sufficient extent through chance perturbations by passing stars. The question as to how the alleged comets got there, lately estimated to be 10^{11} in number—recalling Kepler's view that 'there are as many comets in the sky as fishes in the sea'—and all the additional unnatural requirements that have to be introduced to explain away the abundant difficulties and contradictions that appear as soon as the idea is subjected to test, are dealt with simply by confident assertion that there is evidence tending to support the necessary saving clauses but always without the smallest attempt to say what the evidence is. The deflexion of such comets, even supposing them to exist, by passing stars would be a haphazard process, but the probability that five or six such deflexions should produce a group of sun-grazing comets (which we know to exist) moving in almost identical orbits, is so remote as to be inconceivably small. Of course this by itself does not disprove the speculation, but there is no need to add anything to the advocates' own accounts of the idea to accomplish this.

Another class of hypothesis simply accepts, without explanation, the existence of comets in the solar system and is concerned solely with their structure and development. The coma and the tail are considered to be produced by the warming effect of the sun's radiation as the comet approaches perihelion. Theories of this kind amount to little more than descriptive accounts of a purely *ad hoc* character without providing any real explanation. Indeed, such accounts fail to explain one of the best established features of comets, namely the contraction of the coma as perihelion is approached. According to values computed by Wurm, cometary diameters (D) bear almost a simple inverse relation to the distance from the sun, and he gives the following range of values as typical of cometary dimensions at the corresponding solar distance (r).

TABLE XI

r (a.u.)	D (10^3 km.)
1·5	300–600
1·0	100–300
0·5	20–100
0·3	5–20

This table shows just the reverse of what would be expected if the production of the coma or its changes of size were due to heating effects by the sun. The whole idea of a comet consisting of a gaseous envelope (the coma) retained by a nucleus inevitably fails unless attractive forces of an unknown kind are postulated, since the thermal speeds of particles corresponding to the effective solar temperature at cometary distance are far too high to be controlled by cometary masses, as we have already seen. Any gas developed within the comet would accordingly be lost almost immediately.

11

Where tail-production is concerned, though heating by the sun may be in part the cause, that it cannot be the main factor is shown by the lack of any definite relation between this process and the perihelion distance, q, of the comet. If the release of tail material depended on solar heat, it would be expected that the tails of different comets would always begin to develop at much the same distance from the sun. This, however, is by no means the case. Some comets with values of q greater than $1 \cdot 0$ have developed strong tails while others come much nearer the sun before observable tail-development commences, and yet others with comparatively small values of q show no tails whatever. Then, again, it is difficult to see how the curious activity of Comet 1925 II (Schwassmann-Wachmann), which moves in a nearly circular orbit, could be explained by any action of solar radiation that would not be at least equally applicable to Comet Oterma which also has an almost circular orbit and has shown no similar irregular activity, whereas a dynamical process of the kind here advo-cated obviously contains sufficient flexibility for the occurrence of such activities should the one comet be multiple, for example, and the other a single aggregate of particles.

This, then, gives a general picture of the situation concerning the theory of comets by which the contribution of the present book to the subject is to be assessed. The initial hypotheses of the proposed theory are philosophically on a sound basis, for the existence of dust clouds is an established observational conclusion and therefore a legitimate starting point. The pas-sage of the sun through such clouds is not only a reasonable hypothesis, but the study of the effects of such passages is a theoretical problem automatically arising. As has been seen, the idea involves hypotheses that can be subjected to quanti-tative tests—a feature hitherto completely absent from comet-ary theories—and although these tests are restricted mainly to order of magnitude considerations (as for most cosmogonic hypotheses) the degree of agreement reached is at least as close as could be expected. This might be the moment to recall

162

Jeffreys' percipient remark that most incorrect physical hypotheses, when submitted to order of magnitude test, usually fail by many powers of 10. In the present case it has been seen that satisfactory agreement on this basis emerges at all parts of the theory. There remains little doubt that the theory also explains why so little progress has hitherto been possible with the problem, for the key mechanism is clearly that of accretion, so long overlooked in all branches of astronomy.

The theory confirms what has long been conjectured, namely that planets and comets represent fundamentally different classes of celestial objects. In view of the fact that most if not all stars must from time to time pass through dust clouds, and since a high percentage of stars are double, it is seen that the theory implies that comets probably form the most numerous class of celestial object in the universe. Yet another important consideration emerges. The process of accretion, here applied to interstellar dust, is equally effective for a gaseous cloud in so far as addition of mass to the sun and stars is concerned, and this accretion of mass as stars move through interstellar hydrogen is one of the main factors in determining their growth and course of evolution. The process is of such paramount importance in the theory of stellar evolution that independent supporting evidence for it, especially of an observational kind, is particularly desirable. Such evidence is provided directly by the phenomenon of the comets.

163

APPENDIX: REFERENCES

(Abbreviations: Ap. J.: Astrophysical Journal.
M.N.: Monthly Notices, Royal Astronomical Society.)

The main treatises on comets written during the past century, apart from works dealing entirely with the dynamical motions of comets, include the following:

Hind, J. R., *The Comets*, London, 1852.
Kirkwood, D., *Comets and Meteors*, Philadelphia, 1873.
Guillemin, A., *The World of Comets*, London, 1877.
Shaw, F. G., *Comets and their Tails*, London, 1893.
Chambers, G. F., *The Story of the Comets*, Oxford, 1909.
Olivier, C. P., *Comets*, Baltimore, 1930.
Proctor, Mary, *The Romance of the Comets*, New York and London, 1926.
Proctor, M. and Crommelin, A. C. D., *Comets*, London, 1937.

In addition to these volumes, chapters on comets are to be found in almost all standard treatises on physical astronomy.

Orbits of comets

Where the orbits of comets are concerned, among important papers discussing the cosmogonical aspects of the subject are the following:

Proctor, R. A., *Knowledge*, **vi**, 126, 1887.
Newton, H. A., *Memoirs of Nat. Acad. Sciences*, **6**, 7, 1893.
Russell, H. N., *Astronomical Journal*, **xxxiii**, No. 7, 49, 1920.

The foregoing are on the subject of the capture hypothesis for the short-period comets.

Bredischin, T., *On the Origin of Periodic Comets*, Moscow, 1889.
Callandreau, O., 'Étude sur la Théorie des Comètes Périodiques', *Bull. Soc. Imp. des Naturalistes*, No. 2, 1892.
Forbes, G., *M.N.* **69**, 152, 1909.

The first of these considers other hypotheses for the origin of periodic comets, and the last discusses the possibility of a trans-Neptunian planet inferred from cometary data.

165

The following papers are on the subject of 'hyperbolic' comets:

Fayet, G., *Recherches concernant les Eccentricités des Comètes*, 1906.

Strömgren, E., *Vierteljahrsschrift d. Astr. Gesellschaft*, **iv**, 1910; *Memoirs Acad. Science*, Copenhagen xi, 4, 1914; and *Copenhagen Obs. Publ.* No. 19, 1914, and No. 98, 1935.

van Biesbroeck, G., *Ap. J.* **101**, 376, 1945. ('On the *future* orbit of Comet 1914 V.')

Merton, G., *Observatory*, **67**, 117, 1947. (This article gives a complete list of hyperbolic comets, with references.)

On the distribution of cometary orbits:

Hoek, M., *M.N.* **26**, 147, 1866; and **28**, 129, 1868.

Eddington, A. S., *Observatory*, **36**, No. 459, 142, 1913.

Bobrovnikoff, N. T., *Lick Obs. Bull. XIV*, No. 408, 37, 1929.

Catalogues of the data on cometary orbits:

Galle, J. G., *Verzeichniss der Elemente der bisher Berechneten Cometbahnen*, Leipzig, 1894. (Gives notes on and orbital data of 411 comets from 372 B.C. to 1893.)

Crommelin, A. C. D., *Memoirs of the British Astronomical Association*, **xxvi**, Pt. 2, 1925. (A sequel to Galle's list; contains data of 561 comets.)

Yamamoto, A. S., *Publ. Kwasan Observatory*, i, No. 4, Kyoto, 1936. (Catalogue of comets to 1930.)

Baldet, F., *L'Astronomie*, May, 1929. (Elements of all comets of period less than 170 years.)

Reports on Progress of Astronomy, 'Comets', *M.N.* (Published each year. Full references to all additional calculations of cometary orbits and to other comet literature, compiled at present by G. Merton.)

Cometary spectra

Comprehensive bibliographies on cometary spectra are given in the following articles:

Bobrovnikoff, N. T., *Reviews of Modern Physics*, **14**, 164, 1942.

Swings, P., *M.N.* **105**, 86, 1943.

Halley's comet

N. T. Bobrovnikoff, *Publications of the Lick Observatory*, **xvii**, Part II, 1931. (A comprehensive study of the physical properties of Halley's comet based on observations during the apparition of 1909–1911.)

Brightnesses of comets

Orlov, S. V., *Bulletin of St. Petersburg Acad.* **5**, 1913; and *Publ. of Astro. Soc. of Pacific*, **25**, 175, 1913. (Discusses phase effects in Halley's comet.)

Baldet, F., *Bull. Soc. Astron. de France*, **39**, 577, 1925; *Ann. de l'Obs. d'Astro.*, Paris, vii, 1926; *Trans. I.A.U.*, 155, 1948. (Law of brightness $1/\triangle^2 r^n$.)

Vsessviatsky, S., *Russian Astron. Journal*, **10**, 327, 1933. (Catalogue of brightnesses of 442 comets.)

Bobrovnikoff, N. T., *Perkins Obs. Publ.* **15**, 1941; and **16**, 1942.

Markov, A., *Astro. Nach.* **230**, 151, 1927. (Estimates of masses from brightnesses.)

Structure and dimensions of comets

Richter, N., *Astro. Nach.* **276**, 41, 1948. (Diameters of cometary nuclei.)

Vorontsov-Velyaminov, B. A., *Ap. J.* **104**, 226, 1946. (Masses of cometary nuclei.)

Plummer, H. C., *M.N.* **65**, 237, 1905. (Meteoric constitution of comets.)

Baldet, F., *La Constitution des Comètes*, Paris, 1930.

Wurm, K., *Ap. J.* **89**, 312, 1939.

Dubiago, A. D., *Russian Astron. Journal*, **19**, 14, 1942.

Watson, F. G., *Between the Planets* (Section 5), Toronto, 1941.

Bobrovnikoff, N. T., *Lick Obs. Bull.* No. XIV, 408, 1929. (On the disintegration of comets.)

Minnaert, M. G. J., *Nederlandsche Akad.*, **1**, No. 8, 1947. (On the temperature of cometary nuclei.)

Cometary tails

Bredischin, T., *Mechanische Untersuchungen über Cometenformen*, St. Petersburg, 1903. (Proposes classification of tails into three distinct groups according to their curvatures and lengths, and suggests explanations in terms of light-pressure on different elements.)

Whittaker, E. T., *M.N.* **64**, 347, 1904. (Note on recent researches on the theory of cometary tails.)

Schwarzschild, K., *Ap. J.* **34**, 342, 1911. (On the possible masses of cometary tails.)

Eddington, A. S., *M.N.* **70**, 442, 1910. (On the formation of the envelopes in Morehouse's comet.)

Wurm, K., *Zeits. für Astrophysik*, **10**, 285, 1935. (Discusses selective pressure effects due to resonance and its possible influence on the forms of tails. Also gives numerous references to papers on this topic.)

Radiation pressure effects

Schwarzschild, K., *Sitz. der Math. Phys.* 293, Munich, 1901–2.

Nicholson, J. W., *M.N.* **70**, 544, 1910.

Proudman, J., *M.N.* **73**, 535, 1913.

The above are concerned with the theoretical calculation of the pressure of light on small solid bodies.

Nichols, E. F. and Hull, G. F., 'The Pressure due to Radiation', *Ap. J.* **17**, 315, 1903. (This paper describes the famous experimental proof of light pressure by the sorting of spores from sand in an exhausted enclosure; but the demonstration is incomplete in that the observed motions could also be due to thermal effects.)

Poynting, J. H., *Phil. Trans. Royal Society A*, **202**, 525, 1903; and *Collected Papers*, Cambridge Univ. Press, 1920

Robertson, H. P., *M.N.* **97**, 423, 1937.

These are concerned with the now so-called Poynting-Robertson effect. The second paper gives an accurate relativistic derivation of the appropriate equations of motion.

Plummer, H. C., *M.N.* **65**, 229, 1905, and **66**, 63, 1906. (Considers the application to Encke's comet.)

Crommelin, A. C. D., *M.N.* **80**, 475, 1920. (Shows that there is negligible resistance near the sun to the motions of sun-grazing comets.)

Miscellaneous papers referred to in the text

Eddington, A. S., *M.N.* **76**, 572, 1916. (Distribution of stars in globular clusters.)

Russell, H. N., *Ap. J.* **69**, 49, 1929. (Vaporisation of particles near the sun.)

Beals, C. S., *M.N.* **102**, 96, 1942. (Interstellar dust. Also gives comprehensive references on this subject.)

Hoyle, F., and Bondi, H., *M.N.* **104**, 273, 1944. (Theory of the line accretion problem.)

Parson, A. L., *M.N.* **105**, 244, 1945. (Condensation of dust particles from interstellar gas.)

Hoyle, F., *M.N.* **106**, 406, 1946. (Calculation of heat of evaporation of iron. Condensation of dust particles.)

Jeffreys, H., *M.N.* **107**, 240, 1947. (The effect of cohesion on Roche's limit.)

Kamienski, M., *M.N.* **106**, 267, 1946. (Investigation of Wolf I, deflected into new orbits in 1875 and 1922 by Jupiter.)

Lovell, A. C. B., Porter, J. G., and others, *Reports on Progress in Physics*, **xi**, 389–454, 1948. (Meteors and their relation to comets, with comprehensive references.)

GENERAL INDEX

Acceleration of mean motion, 23, 24, 142

Accretion, 62, 65, 66, 69ff, 100

ACHMAROF, 26

Almagest, 3

ANAXAGORAS, 2

Andromedids, 58

Annual comets, 16, 131

Aphelion, 15, 20, 21, 22, 109, 124

Appearance of comets, 30, 31, 37, 39, 47

Aquarids, 58, 59

ARAGO, 33

ARISTOTLE, 2, 59, 152

Asteroids, 155, 157, 158

BAADE, 51, 155

Babylonians, 2

BACKLUND, 24

BALDET, 46, 52

BARNARD, 25, 26, 30, 31, 33, 36, 44, 45, 62, 125

BESSEL, 4, 12

Bielids, 58, 59

BOBROVNIKOFF, 39, 54, 56, 57, 60

BONDI, 72, 78

BRADLEY, 43

Brightness of comets, 32, 38, 39, 43, 46, 51, 121

BROOKS, 26

Capture hypothesis, 16, 107, 154, 159

CARDAN, 3

CARRINGTON, 159

CASSINI, 54

Centre of gravity of solar system, 98

CHALLIS, 44

CHAMBERLIN, 155, 156

CHAMBERS, 32, 51, 112, 118, 126, 127

Changes of size of comets, 125, 129

Collisions, 83, 120

Colour of comets, 31

Coma, 31, 33, 35, 36, 38, 41, 50, 57, 118

transparency of, 32

Comet groups, 18

Comet-seeker, 24

Comets, *see* Comet Index

CROMMELIN, 18, 26, 56, 59, 60, 156, 157

COWELL, 56

DAIMACA, 26

Daylight comets, 1, 51

Decay of comets, 141

DELAMBRE, 54

DELPORTE, 26

DEMOCRITUS, 2

DENNING, 26

Densities of comets, 48

densities of tails, 50

Designation of comets, 24

Development of comets, 38, 42, 145, 162

Diameters of comets, 40, 41, 51, 162

Discovery of comets, 24

Distribution of particles within comets, 138

Disruption of comets, 19, 42, 45, 49, 113, 149

Diversion of a comet, 134

DOERFEL, 4

DONATI, 55

DUBIAGO, 60

Dust, interstellar, 62

DU TOIT, 26

Eccentricity, 5, 6, 102, 109, 123, 125

Eclipse comets, 24

Ecliptic, 7, 8

EDDINGTON, 14, 15, 21, 34, 138

Emission spectrum, 38, 54, 55

ENCKE, 12

Envelopes surrounding comets, 34, 35, 38, 41, 128

EPHORUS, 44

ESCLANGON, 32

EULER, 12

Expanding shells, or envelopes, 128

FAYET, 11

FEDKE, 26

FORBES, 17, 26

171

COMET INDEX

Biela, 32, 41, 42, 43, 44, 45, 58
Borelly (1903 IV), 36
Brooks (1886), 46
  (1889 V), 13, 44, 45
  (1893), 36
  (1904), 1
  (1911c), 32
  2 (1946e), 24
Brorsen (1846), 17, 24, 43
Chesaux (1744), 36, 52, 54
Coggia (1874 III), 11, 34, 55, 128
Denning (1890 VI), 32
Donati, 11, 23, 32, 34, 128,
Encke, 14, 17, 22, 23, 26, 27, 40, 52,
  54, 58, 112, 117, 118, 119, 122,
  126, 130, 131, 133, 137, 138, 142,
  143, 145
Ensor (1906), 30, 43
Finlay (1893 III), 35
Grigg (1901), 1
Halley, 12, 16, 22, 26, 32, 33, 35, 36,
  38, 39, 40, 41, 44, 45, 50, 52, 54,
  56, 58, 59, 112, 117, 118, 119,
  121, 122, 125, 130, 131, 133, 138
Holmes (1892 III), 17, 30, 31, 33, 40,
  118, 119, 125
Kopff (1906 IV), 44
  (1939e), 27
Lexell (1770), 13, 28, 31, 33, 46
Mellish (1915a), 44
Morehouse (1908 III), 34, 36, 128
Nagato (1931 III), 53
Oterma (1942 VII), 16, 107, 124, 131,
  133, 162
Perrine, 24, 43
Pons-Brooks, 26
Pons-Winnecke, 26, 45, 60, 155
Schwassmann-Wachmann (1925 II),
  10, 16, 26, 107, 124, 125, 131, 162
Stearn (1927 IV), 1, 51
Swift (1899 I), 44
Taylor (1915e), 44
Tempel (1866 I), 24, 25, 43, 55, 58
Tuttle (1871 III), 16
Westphal (1913), 43
Wolf (1933e), 13, 51
Zona (1890 IV), 25

371 B.C., 44
1577, 3, 52
1652, 42
1668, 18
1672, 20
1677, 20
1680, 12, 23
1681, 3
1682, 54
1683, 20
1702, 19
1729, 52
1742, 19
1769, 12
1799 I, 40
1807, 32
1811 I, 52, 118
1812, 19
1823, 33
1825, 36
1843 I, 18, 22, 31, 35
1847 I, 118
1847 V, 118
1849 II, 118
1852, II, 11
1853 III, 105
1860 III, 20, 44
1861 II, 33, 34, 50, 51, 55
1862 III, 58
1863 I, 20
1863 VI, 20, 105
1864, I, 55
1880 I, 18, 19, 31
1882 II, 105
1882 III, 18, 19, 32, 41, 44, 45, 51, 52,
  149
1884 I, 19
1884 III, 19
1886 I, 105
1886 I, 105
1886 II, 105
1886 III, 11
1886 IX, 105
1887 I, 18, 19, 22, 31
1889 I, 105
1890 II, 105
1892 V, 19
1895 II, 13

For EU product safety concerns, contact us at Calle de José Abascal, 56–1°, 28003 Madrid, Spain or eugpsr@cambridge.org.

www.ingramcontent.com/pod-product-compliance
Ingram Content Group UK Ltd.
Pitfield, Milton Keynes, MK11 3LW, UK
UKHW010851090126
466816UK00011B/166